CliffsQuickReview™
Geometry

By Ed Kohn, MS

Hungry Minds™

About the Author

Ed Kohn, MS is an outstanding educator and author with over 33 years experience teaching mathematics. Currently, he is the testing coordinator and math department chairman at Sherman Oaks Center for Enriched Studies.

Publisher's Acknowledgments

Editorial

Project Editor: Sherri Fugit

Acquisitions Editor: Sherry Gomoll

Copy Editor: Robert Annis, Corey Dalton

Technical Editor: David Herzog, Michael Kantor

Editorial Assistant: Jennifer Young

Special Help

David Herzog, Carol Strickland, Jennifer Young

Production

Proofreader: Betty Kish

Indexer: TECHBOOKS Production Services

Hungry Minds Indianapolis Production Services

CliffsQuickReview™ Geometry

Published by
Hungry Minds, Inc.
909 Third Avenue
New York, NY 10022
www.hungryminds.com
www.cliffsnotes.com

Library of Congress Control Number: 2001024220

ISBN: 0-7645-6380-7

Printed in the United States of America

10 9 8

1O/QV/QT/QS/IN

Distributed in the United States by Hungry Minds, Inc.

Distributed by CDG Books Canada Inc. for Canada; by Transworld Publishers Limited in the United Kingdom; by IDG Norge Books for Norway; by IDG Sweden Books for Sweden; by IDG Books Australia Publishing Corporation Pty. Ltd. for Australia and New Zealand; by TransQuest Publishers Pte Ltd. for Singapore, Malaysia, Thailand, Indonesia, and Hong Kong; by Gotop Information Inc. for Taiwan; by ICG Muse, Inc. for Japan; by Intersoft for South Africa; by Eyrolles for France; by International Thomson Publishing for Germany, Austria and Switzerland; by Distribuidora Cuspide for Argentina; by LR International for Brazil; by Galileo Libros for Chile; by Ediciones ZETA S.C.R. Ltda. for Peru; by WS Computer Publishing Corporation, Inc., for the Philippines; by Contemporanea de Ediciones for Venezuela; by Express Computer Distributors for the Caribbean and West Indies; by Micronesia Media Distributor, Inc. for Micronesia; by Chips Computadoras S.A. de C.V. for Mexico; by Editorial Norma de Panama S.A. for Panama; by American Bookshops for Finland.

For general information on Hungry Minds' products and services in the U.S., please call our Customer Care department at 800-762-2974.

For reseller information, including discounts and premium sales, please call our Reseller Customer Care department at 800-434-3422.

For information on where to purchase Hungry Minds' products outside the U.S., please contact our International Sales department at 317-572-3993 or fax 317-572-4002.

For consumer information on foreign language translations, please contact our Customer Care department at 800-434-3422, fax 317-572-4002, or e-mail rights@hungryminds.com.

For information on licensing foreign or domestic rights, please contact our Sub-Rights department at 212-884-5000.

For sales inquiries and special prices for bulk quantities, please contact our Order Services department at 800-434-3422 or write to Hungry Minds, Inc., Attn: Customer Care department, 10475 Crosspoint Boulevard, Indianapolis, IN 46256.

For information on using Hungry Minds' products and services in the classroom or for ordering examination copies, please contact our Educational Sales department at 800-434-2086 or fax 317-572-4005.

Please contact our Public Relations department at 212-884-5163 for press review copies or 212-884-5000 for author interviews and other publicity information or fax 212-884-5400.

For authorization to photocopy items for corporate, personal, or educational use, please contact Copyright Clearance Center, 222 Rosewood Drive, Danvers, MA 01923, or fax 978-750-4470.

Table of Contents

Introduction

Learning geometry has traditionally been regarded as important in the secondary schools, at least partly because it has been the primary means of teaching the art of reasoning. Over time, geometry has evolved into a beautifully arranged and logically organized body of knowledge. It consists primarily of a sequence of statements about points, lines, and planes, as well as planar and solid figures. Geometry begins with undefined terms, definitions, and assumptions; these lead to theorems and constructions. It is an abstract subject, but easy to visualize, and it has many concrete practical applications. Historically, geometry has long been important for its role in the surveying of land and more recently, our knowledge of geometry has been applied to help build structurally sound bridges, experimental space stations, and large athletic and entertainment arenas, just to mention a few examples.

Unfortunately, many students planning to take standardized college admission tests or to pursue careers requiring more advanced mathematics and physics overlook the significant role of geometry in these undertakings. This book helps you get ready for success on the mathematics portion of these standardized tests or in the more advanced mathematics and science courses required for careers in mathematics, the sciences, and engineering. Teachers and students alike will find CQR *Geometry* to be a valuable course supplement.

The prerequisites for subject comprehension include familiarity with basic arithmetic operations and with some topics from introductory algebra. The arithmetic prerequisites include converting fractions to decimals, converting improper fractions to mixed numbers, and taking square roots. The algebraic prerequisites include addition of equations, solving linear equations in two variables for one variable in terms of the other, and factoring quadratic equations.

Why You Need This Book

Can you answer yes to any of these questions?

- Do you need to review the fundamentals of geometry fast?
- Do you need a course supplement to geometry?
- Do you need to prepare for your geometry test?
- Do you need a concise, comprehensive reference for geometry?

If so, then Cliffs Quick Review *Geometry* is for you!

How to Use This Book

Because mathematics builds on itself, many readers benefit most from studying or reviewing this book from cover to cover. However, you're the boss here, and you may choose to seek only the information you want and then put the book back on the shelf for later. In that case, here are a few recommended ways to search for topics:

■ Flip through the book, looking for your topic in the running heads.

■ Look in the Glossary for all the important terms and definitions.

■ Look for your topic in the Table of Contents in the front of the book.

■ Look at the In This Chapter section at the front of each chapter.

■ Look at the review questions in the Chapter Checkout.

■ Look for additional information in the CQR Resource Center or test your knowledge with the CQR Review.

■ Refer to the CQR Pocket Guide.

■ Flip through the book and look at the many figures, reading the captions until you find what you're looking for.

Don't Miss Our Web Site

Keep up with the latest resources available by visiting the CliffsNotes web site at www.cliffsnotes.com. Here's what you find:

■ Informative interactive tools

■ Links to related web sites

■ Additional resources to help you continue your learning

At www.cliffsnotes.com, you can even register for a new feature called CliffsNotes Daily, which offers you newsletters on a variety of topics and is delivered right to your e-mail inbox each business day.

If you haven't yet discovered the Internet and are wondering how to get online, pick up *Getting On the Internet,* new from CliffsNotes. You'll learn just what you need to make your online connection quickly and easily. See you at www.cliffsnotes.com!

Chapter 1

FUNDAMENTAL IDEAS

Chapter Check-In

❏ Understanding what is meant by point, line, and plane

❏ Knowing the relationship between postulates and theorems

❏ Computing the midpoint of a line segment

❏ Identifying acute, right, obtuse, and straight angles as well as complementary angles, supplementary angles, and vertical angles

❏ Identifying parallel lines and perpendicular lines

Geometry was the first system of ideas in which a few simple statements were assumed and then used to derive more complex ones. A system such as this is referred to as a deductive system. Geometry introduces you to the ideas of deduction and logical consequences, ideas you will continue to use throughout your life.

Points, Lines, and Planes

Point, line, and *plane,* together with *set,* are the undefined terms that provide the starting place for geometry. When we define words, we ordinarily use simpler words, and these simpler words are in turn defined using yet simpler words. This process must eventually terminate; at some stage, the definition must use a word whose meaning is accepted as intuitively clear. Because that meaning is accepted without definition, we refer to these words as *undefined terms.* These terms will be used in defining other terms. Although these terms are not formally defined, a brief intuitive discussion is needed.

Point

A **point** is the most fundamental object in geometry. It is represented by a dot and named by a capital letter. A point represents position only; it has zero size (that is, zero length, zero width, and zero height). Figure 1-1 illustrates point *C*, point *M*, and point *Q*.

Figure 1-1   Three points.

Line

A **line** *(straight line)* can be thought of as a connected set of infinitely many points. It extends infinitely far in two opposite directions. A line has infinite length, zero width, and zero height. Any two points on the line name it. The symbol ↔ written on top of two letters is used to denote that line. A line may also be named by one small letter (Figure 1-2).

Figure 1-2   Two lines.

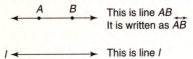

Collinear points

Points that lie on the same line are called **collinear points.** If there is no line on which all of the points lie, then they are **noncollinear points.** In Figure 1-3, points *M, A,* and *N* are collinear, and points *T, I,* and *C* are noncollinear.

Figure 1-3   Three collinear points and three noncollinear points.

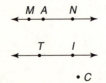

Plane

A **plane** may be considered as an infinite set of points forming a connected flat surface extending infinitely far in all directions. A plane has infinite length, infinite width, and zero height (or thickness). It is usually represented in drawings by a four-sided figure. A single capital letter is used to denote a plane. The word *plane* is written with the letter so as not to be confused with a point (Figure 1-4).

Figure 1-4 Two planes.

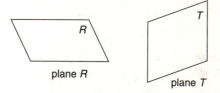

plane *R*

plane *T*

Postulates and Theorems

As mentioned at the beginning of this chapter, a postulate is a statement that is assumed true without proof. A theorem is a true statement that can be proven. Listed below are the first six postulates and the theorems that can be proven from these postulates.

Postulate 1: A line contains at least two points.

Postulate 2: A plane contains at least three noncollinear points.

Postulate 3: Through any two points, there is exactly one line.

Postulate 4: Through any three noncollinear points, there is exactly one plane.

Postulate 5: If two points lie in a plane, then the line joining them lies in that plane.

Postulate 6: If two planes intersect, then their intersection is a line.

Theorem 1: If two lines intersect, then they intersect in exactly one point.

Theorem 2: If a point lies outside a line, then exactly one plane contains both the line and the point.

Theorem 3: If two lines intersect, then exactly one plane contains both lines.

Example 1: State the postulate or theorem you would use to justify the statement made about each figure.

Figure 1-5 Illustrations of Postulates 1–6 and Theorems 1–3.

One plane contains points *A*, *B*, and *C*.

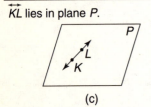

(a)

Only one line contains points *Q* and *T*.

Q.

• *T*

(b)

$\overleftrightarrow{KL}$ lies in plane *P*.

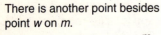

(c)

Plane *G* and plane *H* intersect along line *l*.

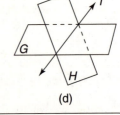

(d)

There is another point besides point *w* on *m*.

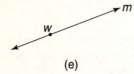

(e)

One plane contains *t* and *l*.

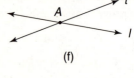

(f)

One plane contains $\overleftrightarrow{AC}$ and *B*.

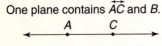

(g)

Lines *q* and *r* intersect at *M* and at no other point.

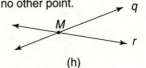

(h)

(a) Through any three noncollinear points, there is exactly one plane *(Postulate 4)*.

(b) Through any two points, there is exactly one line *(Postulate 3)*.

(c)   If two points lie in a plane, then the line joining them lies in that plane *(Postulate 5)*.

(d)   If two planes intersect, then their intersection is a line *(Postulate 6)*.

(e)   A line contains at least two points *(Postulate 1)*.

(f)   If two lines intersect, then exactly one plane contains both lines *(Theorem 3)*.

(g)   If a point lies outside a line, then exactly one plane contains both the line and the point *(Theorem 2)*.

(h)   If two lines intersect, then they intersect in exactly one point *(Theorem 1)*.

Segments, Midpoints, and Rays

The concept of lines has already been introduced, but much of geometry is concerned with portions of lines. Some of those portions are so special that they have their own names and symbols.

Line segment

A line segment is a connected piece of a line. It has two endpoints and is named by its endpoints. Sometimes, the symbol – written on top of two letters is used to denote the segment. This is line segment *CD* (Figure 1-6).

Figure 1-6   Line segment.

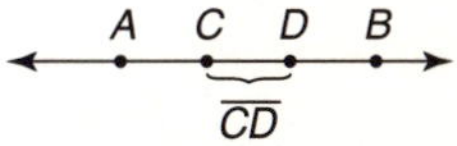

$$A \quad C \quad D \quad B$$
$$\overline{CD}$$

It is written $\overline{CD}$. (Technically, $\overline{CD}$ refers to the points C and D and all the points between them, and $\overleftrightarrow{CD}$ without the ‾ refers to the distance from C to D.) Note that $\overline{CD}$ is a piece of $\overleftrightarrow{AB}$.

Postulate 7 (Ruler Postulate): Each point on a line can be paired with exactly one real number called its **coordinate.** The distance between two points is the positive difference of their coordinates (Figure 1-7).

Figure 1-7   Distance between two points.

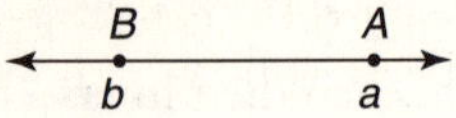

If $a > b$, then $AB = a - b$.

Example 2: In Figure 1-8, find the length of QU.

Figure 1-8   Length of a line segment.

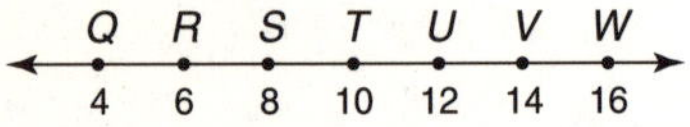

$QU = 12 - 4$

$QU = 8$ (The length of $\overline{QU}$ is 8.)

Postulate 8 (Segment Addition Postulate): If B lies between A and C on a line, then $AB + BC = AC$ (Figure 1-9).

Figure 1-9   Addition of lengths of line segments.

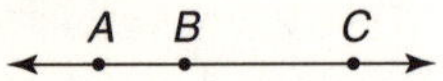

Example 3: In Figure 1-10, A lies between C and T. Find CT if $CA = 5$ and $AT = 8$.

Figure 1-10   Addition of lengths of line segments.

Because A lies between C and T, Postulate 8 tells you

$$CA + AT = CT$$
$$5 + 8 = 13$$
$$CT = 13$$

Midpoint

A **midpoint** of a line segment is the halfway point, or the point equidistant from the endpoints (Figure 1-11).

Figure 1-11   Midpoint of a line segment.

$$Q \qquad R \qquad S$$

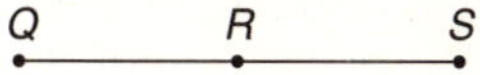

R is the midpoint of $\overline{QS}$ because $QR = RS$ or because $QR = \frac{1}{2}QS$ or $RS = \frac{1}{2}QS$.

Example 4: In Figure 1-12, find the midpoint of $\overline{KR}$.

Figure 1-12   Midpoint of a line segment.

K	L	M	N	O	P	Q	R
5	8	11	12	17	19	23	29

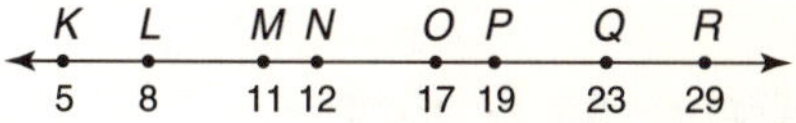

$$KR = 29 - 5 \text{ or } KR = 24$$

The midpoint of $\overline{KR}$ would be $\frac{1}{2}(24)$ or 12 spaces from either *K* or *R*. Because the coordinate of *K* is 5, and it is smaller than the coordinate of R (which is 29), to get the coordinate of the midpoint you could either add 12 to 5 or subtract 12 from 29. In either case, you determine that the coordinate of the midpoint is 17. That means that point *O* is the midpoint of $\overline{KR}$ because $KO = OR$.

Another way to get the coordinate of the midpoint would be to find the average of the endpoint coordinates. To find the average of two numbers, you find their sum and divide by two. $(5 + 29) \div 2 = 17$. The coordinate of the midpoint is 17, so the midpoint is point *O*.

Theorem 4: A line segment has exactly one midpoint.

Ray

A **ray** is also a piece of a line, except that it has only one endpoint and continues forever in one direction. It could be thought of as a half-line with an endpoint. It is named by the letter of its endpoint and any other point on the ray. The symbol → written on top of the two letters is used to denote that ray. This is ray *AB* (Figure 1-13).

Figure 1-13 Ray *AB*.

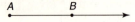

It is written as $\overrightarrow{AB}$.

This is ray *CD* (Figure 1-14).

Figure 1-14 Ray *CD*.

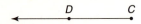

It is written as $\overrightarrow{CD}$ or, $\overleftarrow{DC}$

Note that the nonarrow part of the ray symbol is over the endpoint.

Angles and Angle Pairs

Easily as significant as rays and line segments are the angles they form. Without them, there would be none of the geometric figures that you know (with the possible exception of the circle).

Angles

Two rays that have the same endpoint form an angle. That endpoint is called the vertex, and the rays are called the **sides** of the angle. In geometry, an angle is measured in **degrees** from 0° to 180°. The number of degrees indicates the size of the angle. In Figure 1-15, rays $\overrightarrow{AB}$ and $\overrightarrow{AC}$ form the angle. *A* is the vertex. $\overrightarrow{AB}$ and $\overrightarrow{AC}$ are the sides of the angle.

Figure 1-15 ∠*BAC*.

The symbol ∠ is used to denote an angle. The symbol $m \angle$ is sometimes used to denote the measure of an angle.

An angle can be named in various ways (Figure 1-16).

Figure 1-16 Different names for the same angle.

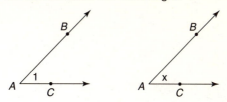

- By the letter of the vertex—therefore, the angle in Figure 1-16 could be named ∠*A*.

- By the number (or small letter) in its interior—therefore, the angle in Figure 1-16 could be named ∠1 or ∠*x*.

- By the letters of three points that form it—therefore, the angle in Figure 1-16 could be named ∠*BAC* or ∠*CAB*. The center letter is always the letter of the vertex.

Example 5: In Figure 1-17 (a) use three letters to rename ∠3; (b) use one number to rename ∠*KMJ*.

Figure 1-17 Different names for the same angle.

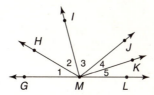

(a) ∠3 is the same as ∠*IMJ* or ∠*JMI*;

(b) ∠*KMJ* is the same as ∠4.

Postulate 9 (Protractor Postulate): Suppose *O* is a point on $\overleftrightarrow{XY}$. Consider all rays with endpoint *O* that lie on one side of $\overleftrightarrow{XY}$. Each ray can be paired with exactly one real number between 0° and 180°, as shown in Figure 1-18. The positive difference between two numbers representing two different rays is the measure of the angle whose sides are the two rays.

Figure 1-18 Using the Protractor Postulate.

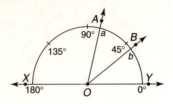

If *a* > *b*, then *m* ∠*AOB* = *a* - *b*.

Example 6: Use Figure 1-19 to find the following: (a) *m* ∠*SON,* (b) *m* ∠*ROT,* and (c) *m* ∠*MOE.*

Figure 1-19 Using the Protractor Postulate.

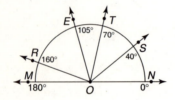

(a) *m* ∠*SON* = 40° - 0°

  m ∠*SON* = 40°

(b) *m* ∠*ROT* = 160° - 70°

  m ∠*ROT* = 90°

(c) *m* ∠*MOE* = 180° - 105°

  m ∠*MOE* = 75°

Postulate 10 (Angle Addition Postulate): If $\overrightarrow{OB}$ lies between $\overrightarrow{OA}$ and $\overrightarrow{OC}$, then *m* ∠*AOB* + *m* ∠*BOC* = *m* ∠*AOC* (Figure 1-20).

Figure 1-20 Addition of angles.

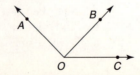

Example 7: In Figure 1-21, if $m \angle 1 = 32°$ and $m \angle 2 = 45°$, find $m \angle NEC$.

Figure 1-21   Addition of angles.

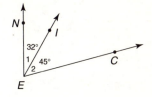

Because $\overrightarrow{EI}$ is between $\overrightarrow{EN}$ and $\overrightarrow{EC}$, by *Postulate 10*,

$$m \angle 1 + m \angle 2 = m \angle NEC$$
$$32° + 45° = m \angle NEC$$
$$m \angle NEC = 77°$$

Angle bisector

An **angle bisector** is a ray that divides an angle into two equal angles. In Figure 1-22 $\overrightarrow{OY}$ is a bisector of $\angle XOZ$ because $= m \angle XOY = m \angle YOZ$.

Figure 1-22   Bisector of an angle.

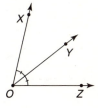

Theorem 5: An angle that is not a straight angle has exactly one bisector.

Certain angles are given special names based on their measures.

Right angle

A **right angle** has a measure of 90°. The symbol ⌐ in the interior of an angle designates the fact that a right angle is formed. In Figure 1-23, $\angle ABC$ is a right angle.

Figure 1-23 A right angle.

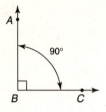

$$m \angle ABC = 90°$$

Theorem 6: All right angles are equal.

Acute angle

An **acute angle** is any angle whose measure is less than 90°. In Figure 1-24, $\angle b$ is acute.

Figure 1-24 An acute angle.

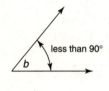

$$m \angle b < 90°$$

Obtuse angle

An **obtuse angle** is an angle whose measure is more than 90° but less than 180°. In Figure 1-25, $\angle 4$ is obtuse.

Figure 1-25 An obtuse angle.

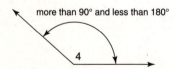

$$m \angle 4 > 90° \text{ and } m \angle 4 < 180°$$

or

$$90° < m \angle 4 < 180°$$

Straight angle

Some geometry texts refer to an angle with a measure of 180° as a **straight angle.** In Figure 1-26, $\angle BAC$ is a straight angle.

Figure 1-26   A straight angle.

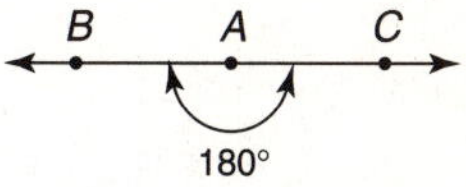

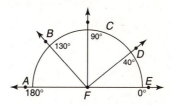

$$m \angle BAC = 180°$$

Example 8: Use Figure 1-27 to identify each named angle as acute, right, obtuse, or straight: (a) $\angle BFD$, (b) $\angle AFE$, (c) $\angle BFC$, (d) $\angle DFA$.

Figure 1-27   Classification of angles.

(a) $m \angle BFD = 90°$ (130° - 40° = 90°), so $\angle BFD$ is a right angle.

(b) $m \angle AFE = 180°$, so $\angle AFE$ is a straight angle.

(c) $m \angle BFC = 40°$ (130° - 90° = 40°), so $\angle BFC$ is an acute angle.

(d) $m \angle DFA = 140°$ (180° - 40° = 140°), so $\angle DFA$ is an obtuse angle.

Special Angles

Certain angle pairs are given special names based on their relative position to one another or based on the sum of their respective measures.

Adjacent angles

Adjacent angles are any two angles that share a common side separating the two angles and that share a common vertex. In Figure 1-28, ∠1 and ∠2 are adjacent angles.

Figure 1-28 Adjacent angles.

Vertical angles

Vertical angles are formed when two lines intersect and form four angles. Any two of these angles that are *not* adjacent angles are called vertical angles. In Figure 1-29, line *l* and line *m* intersect at point *Q*, forming ∠1, ∠2, ∠3, and ∠4.

Figure 1-29 Two pairs of vertical angles and four pairs of adjacent angles.

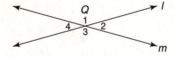

- Vertical angles:

    ∠1 and ∠3

    ∠2 and ∠4

- Adjacent angles:

    ∠1 and ∠2

    ∠2 and ∠3

    ∠3 and ∠4

    ∠4 and ∠1

Theorem 7: Vertical angles are equal in measure.

Complementary angles

Complementary angles are any two angles whose sum is 90°. In Figure 1-30, because $\angle ABC$ is a right angle, $m \angle 1 + m \angle 2 = 90°$, so $\angle 1$ and $\angle 2$ are complementary.

Figure 1-30   Adjacent complementary angles.

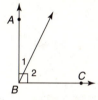

Complementary angles do not need to be adjacent. In Figure 1-31, because $m \angle 3 + m \angle 4 = 90°$, $\angle 3$, and $\angle 4$, are complementary.

Figure 1-31   Nonadjacent complementary angles.

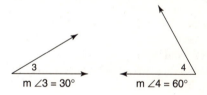

$$m \angle 3 = 30° \qquad m \angle 4 = 60°$$

Example 9: If $\angle 5$ and $\angle 6$ are complementary, and $m \angle 5 = 15°$, find $m \angle 6$.

Because $\angle 5$ and $\angle 6$ are complementary,

$$m \angle 5 + m \angle 6 = 90°$$
$$15 + m \angle 6 = 90°$$
$$m \angle 6 = 90° - 15°$$
$$m \angle 6 = 75°$$

Theorem 8: If two angles are complementary to the same angle, or to equal angles, then they are equal to each other.

Refer to Figures 1-32 and 1-33. In Figure 1-32, $\angle A$ and $\angle B$ are complementary. Also, $\angle C$ and $\angle B$ are complementary. *Theorem 8* tells you that

$m \angle A = m \angle C$. In Figure 1-33, $\angle A$ and $\angle B$ are complementary. Also, $\angle C$ and $\angle D$ are complementary, and $m \angle B = m \angle D$. *Theorem 8* now tells you that $m \angle A = m \angle C$.

Figure 1-32 Two angles complementary to the same angle.

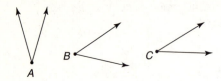

Figure 1-33 Two angles complementary to equal angles.

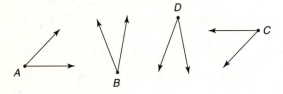

Supplementary angles

Supplementary angles are two angles whose sum is 180°. In Figure 1-34, $\angle ABC$ is a straight angle. Therefore $m \angle 6 + m \angle 7 = 180°$, so $\angle 6$ and $\angle 7$ are supplementary.

Figure 1-34 Adjacent supplementary angles.

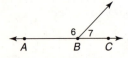

Theorem 9: If two adjacent angles have their noncommon sides lying on a line, then they are supplementary angles.

Supplementary angles do not need to be adjacent (Figure 1-35).

Figure 1-35   Nonadjacent supplementary angles.

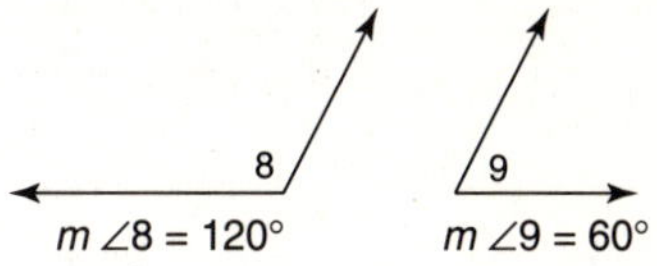

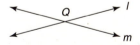

Because $m \angle 8 + m \angle 9 = 180°$, $\angle 8$ and $\angle 9$ are supplementary.

Theorem 10: If two angles are supplementary to the same angle, or to equal angles, then they are equal to each other.

Intersecting, Perpendicular, and Parallel Lines

You have probably had the experience of standing in line for a movie ticket, a bus ride, or something for which the demand was so great it was necessary to wait your turn. However, in geometry, there are three types of lines that students should understand.

Intersecting lines

Two or more lines that meet at a point are called *intersecting lines*. That point would be on each of these lines. In Figure 1-36, lines *l* and *m* intersect at *Q*.

Figure 1-36   Intersecting lines.

Perpendicular lines

Two lines that intersect and form right angles are called **perpendicular lines.** The symbol ⊥ is used to denote perpendicular lines. In Figure 1-37, line *l* ⊥ line *m*.

Figure 1-37 Perpendicular lines.

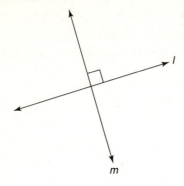

Parallel lines

Two lines, both in the same plane, that never intersect are called **parallel lines.** Parallel lines remain the same distance apart at all times. The symbol // is used to denote parallel lines. In Figure 1-38, *l // m*..

Figure 1-38 Parallel lines.

Parallel and Perpendicular Planes

You may be tempted to think of planes as vehicles to be found up in the sky or at the airport. Well, rest assured, geometry is no fly-by-night operation.

Parallel planes

Parallel planes are two planes that do not intersect. In Figure 1-39, plane *P //* plane *Q.*

Figure 1-39 Parallel planes.

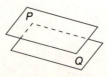

Theorem 11: If each of two planes is parallel to a third plane, then the two planes are parallel to each other (Figure 1-40).

Figure 1-40 Two planes parallel to a third plane.

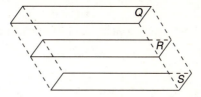

plane *Q* // plane *R*, plane *S* // plane *R*; thus, plane *Q* // plane *S*

Perpendicular planes

A line *l* is perpendicular to plane *A* if *l* is perpendicular to all of the lines in plane *A* that intersect *l*. (Think of a stick standing straight up on a level surface. The stick is perpendicular to all of the lines drawn on the table that pass through the point where the stick is standing).

A plane *B* is perpendicular to a plane *A* if plane *B* contains a line that is perpendicular to plane *A*. (Think of a book balanced upright on a level surface.) See Figure 1-41.

Figure 1-41 Perpendicular planes.

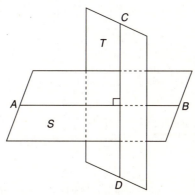

Theorem 12: If two planes are perpendicular to the same plane, then the two planes either intersect or are parallel.

In Figure 1-42, plane $B \perp$ plane A, plane $C \perp$ plane A, and plane B and plane C intersect along line l.

Figure 1-42 Two intersecting planes that are perpendicular to the same plane.

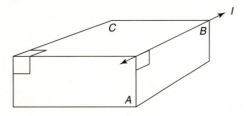

In Figure 1-43, plane $B \perp$ plane A, plane $C \perp$ plane A, and plane B // plane C.

Figure 1-43 Two parallel planes that are perpendicular to the same plane.

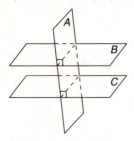

Chapter Checkout

Q&A

1. True or False: A postulate is a mathematical statement that has been proved.
2. If $A = 11$ and $B = 23$, find the midpoint of AB.

A transversal that intersects two lines forms eight angles; certain pairs of these angles are given special names. They are as follows:

- **Corresponding angles** are the angles that appear to be in the same relative position in each group of four angles. In Figure 2-2, ∠1 and ∠5 are corresponding angles. Other pairs of corresponding angles in Figure 2-2 are: ∠4 and ∠8, ∠2 and ∠6, and ∠3 and ∠7.

Figure 2-2 A transveral intersecting two lines and forming various pairs of corresponding angles—alternate interior angles, alternate exterior angles, consecutive interior angles, and consecutive exterior angles.

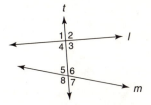

- **Alternate interior angles** are angles within the lines being intersected, on opposite sides of the transversal, and are not adjacent. In Figure 2-2, ∠4 and ∠6 are alternate interior angles. Also, ∠3 and ∠5 are alternate interior angles.

- **Alternate exterior angles** are angles outside the lines being intersected, on opposite sides of the transversal, and are not adjacent. In Figure 2-2, ∠1 and ∠7 are alternate exterior angles. Also, ∠2 and ∠8 are alternate exterior angles.

- **Consecutive interior angles** (same-side interior angles) are interior angles on the same side of the transversal. In Figure 2-2, ∠4 and ∠5 are consecutive interior angles. Also, ∠3 and ∠6 are consecutive interior angles.

- **Consecutive exterior angles** (same-side exterior angles) are exterior angles on the same side of the transversal. In Figure 2-2, ∠1 and ∠8 are consecutive exterior angles. Also, ∠2 and ∠7 are consecutive exterior angles.

The Parallel Postulate

Postulate 11 (Parallel Postulate): If two parallel lines are cut by a transversal, then the corresponding angles are equal (Figure 2-3).

Figure 2-3 Corresponding angles are equal when two parallel lines are cut by a transversal.

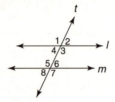

This postulate says that if *l // m,* then

- $m \angle 1 = m \angle 5$
- $m \angle 2 = m \angle 6$
- $m \angle 3 = m \angle 7$
- $m \angle 4 = m \angle 8$

Consequences of the Parallel Postulate

Postulate 11 can be used to derive additional theorems regarding parallel lines cut by a transversal. In Figure 2-3, because $m \angle 1 + m \angle 2 = 180°$ and $m \angle 5 + m \angle 6 = 180°$ (because adjacent angles whose noncommon sides lie on a line are supplementary), and because $m \angle 1 = m \angle 3$, $m\angle 2 = m \angle 4$, $m \angle 5 = m \angle 7$, and $m \angle 6 = m \angle 8$ (because vertical angles are equal), all of the following theorems can be proven as a consequence of *Postulate 11.*

Theorem 13: If two parallel lines are cut by a transversal, then alternate interior angles are equal.

Theorem 14: If two parallel lines are cut by a transversal, then alternate exterior angles are equal.

Theorem 15: If two parallel lines are cut by a transversal, then consecutive interior angles are supplementary.

Theorem 16: If two parallel lines are cut by a transversal, then consecutive exterior angles are supplementary.

The above postulate and theorems can be condensed to the following theorems:

Theorem 17: If two parallel lines are cut by a transversal, then every pair of angles formed are either equal or supplementary.

Theorem 18: If a transversal is perpendicular to one of two parallel lines, then it is also perpendicular to the other line.

Based on *Postulate 11* and the theorems that follow it, all of the following conditions would be true if *l // m* (Figure 2-4).

Figure 2-4   Two parallel lines cut by a transversal.

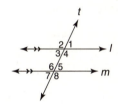

In figures, single or double arrows on a pair of lines indicate that the lines are parallel.

Based on *Postulate 11:*

■ $m \angle 1 = m \angle 5$

■ $m \angle 4 = m \angle 8$

■ $m \angle 2 = m \angle 6$

■ $m \angle 3 = m \angle 7$

Based on *Theorem 13:*

■ $m \angle 3 = m \angle 5$

■ $m \angle 4 = m \angle 6$

Based on *Theorem 14:*

■ $m \angle 1 = m \angle 7$

■ $m \angle 2 = m \angle 8$

Based on *Theorem 15:*

- ∠3 and ∠6 are supplementary

- ∠4 and ∠5 are supplementary

Based on *Theorem 16:*

- ∠1 and ∠8 are supplementary

- ∠2 and ∠7 are supplementary

Based on *Theorem 18:*

- If $t \perp l$, then $t \perp m$

Testing for Parallel Lines

Postulate 11 and Theorems 13 through 18 tell you that *if* two lines are parallel, *then* certain other statements are also true. It is often useful to show that two lines are in fact parallel. For this purpose, you need theorems in the following form: *If* (certain statements are true) *then* (two lines are parallel). It is important to realize that the **converse** of a theorem (the statement obtained by switching the *if* and *then* parts) is not always true. In this case, however, the converse of postulate 11 turns out to be true. We state the converse of Postulate 11 as Postulate 12 and use it to prove that the converses of Theorems 13 through 18 are also theorems.

Postulate 12: If two lines and a transversal form equal corresponding angles, then the lines are parallel.

In Figure 2-5, if $m \angle 1 = m \angle 2$, then $l \parallel m$. (Any pair of equal corresponding angles would make $l \parallel m$.)

Figure 2-5 A transversal cuts two lines to form equal corresponding angles.

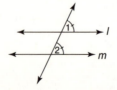

This postulate allows you to prove that all the converses of the previous theorems are also true.

Theorem 19: If two lines and a transversal form equal alternate interior angles, then the lines are parallel.

Theorem 20: If two lines and a transversal form equal alternate exterior angles, then the lines are parallel.

Theorem 21: If two lines and a transversal form consecutive interior angles that are supplementary, then the lines are parallel.

Theorem 22: If two lines and a transversal form consecutive exterior angles that are supplementary, then the lines are parallel.

Theorem 23: In a plane, if two lines are parallel to a third line, the two lines are parallel to each other.

Theorem 24: In a plane, if two lines are perpendicular to the same line, then the two lines are parallel.

Based on *Postulate 12* and the theorems that follow it, any of following conditions would allow you to prove that *a // b*. (Figure 2-6).

Figure 2-6   What conditions on these numbered angles would guarantee that lines *a* and *b* are parallel?

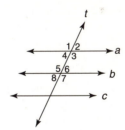

Use *Postulate 12:*

- $m \angle 1 = m \angle 5$
- $m \angle 2 = m \angle 6$
- $m \angle 3 = m \angle 7$
- $m \angle 4 = m \angle 8$

Use *Theorem 19:*

■ $m \angle 4 = m \angle 6$

■ $m \angle 3 = m \angle 5$

Use *Theorem 20:*

■ $m \angle 1 = m \angle 7$

■ $m \angle 2 = m \angle 8$

Use *Theorem 21:*

■ $\angle 4$ and $\angle 5$ are supplementary

■ $\angle 3$ and $\angle 6$ are supplementary

Use *Theorem 22:*

■ $\angle 1$ and $\angle 8$ are supplementary

■ $\angle 2$ and $\angle 7$ are supplementary

Use *Theorem 23:*

■ $a \parallel c$ and $b \parallel c$

Use *Theorem 24:*

■ $a \perp t$ and $b \perp t$

Example 1: Using Figure 2-7, identify the given angle pairs as alternate interior, alternate exterior, consecutive interior, consecutive exterior, corresponding, or none of these: $\angle 1$ and $\angle 7$, $\angle 2$ and $\angle 8$, $\angle 3$ and $\angle 4$, $\angle 4$ and $\angle 8$, $\angle 3$ and $\angle 8$, $\angle 3$, and $\angle 2$, $\angle 5$ and $\angle 7$.

Figure 2-7 Find the angle pairs that are alternate interior, alternate exterior, consecutive interior, consecutive exterior, and corresponding.

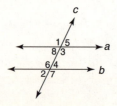

∠1 and ∠7 are alternate exterior angles.

∠2 and ∠8 are corresponding angles.

∠3 and ∠4 are consecutive interior angles.

∠4 and ∠8 are alternate interior angles.

∠3 and ∠2 are none of these.

∠5 and ∠7 are consecutive exterior angles.

Example 2: For each of the figures in Figure 2-8, determine which postulate or theorem you would use to prove $l \parallel m$.

Figure 2-8 Conditions guaranteeing that lines l and m are parallel.

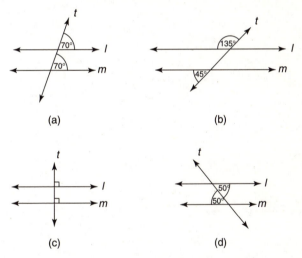

(a)

(b)

(c)

(d)

Figure 2-8(a): If two lines and a transversal form equal corresponding angles, then the lines are parallel *(Postulate 12)*.

Figure 2-8(b): If two lines and a transversal form consecutive exterior angles that are supplementary, then the lines are parallel *(Theorem 22)*.

Figure 2-8(c): In a plane, if two lines are perpendicular to the same line, the two lines are parallel *(Theorem 24)*.

Figure 2-8(d): If two lines and a transversal form equal alternate interior angles, then the lines are parallel *(Theorem 19)*.

Example 3: In Figure 2-9, $a \parallel b$ and $m\angle 1 = 117°$. Find the measure of each of the numbered angles.

Figure 2-9 When lines *a* and *b* are parallel, knowing one angle makes it possible to determine all the others pictured here.

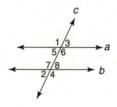

$m\angle 2 = 63°$

$m\angle 3 = 63°$

$m\angle 4 = 117°$

$m\angle 5 = 63°$

$m\angle 6 = 117°$

$m\angle 7 = 117°$

$m\angle 8 = 63°$

Chapter Checkout

Q&A

1. When two parallel lines are cut by a transversal, what is true about corresponding angles?

2. Refer to Figure 2-2. If $m\angle 4 + m\angle 5 = 180°$, then what does that guarantee about lines *l* and *m*?

3. Refer to Figure 2-2. If $m\angle 1 = m\angle 5$, then what does that guarantee about lines *l* and *m*?

4. $\overleftrightarrow{AB}$, $\overleftrightarrow{CD}$ and $\overleftrightarrow{EF}$ are parallel to one another and $m\angle A = 39°$ and $m\angle E = 65°$ in Figure 2-10. Find $m\angle ACE$.

Figure 2-10 These lines form angles.

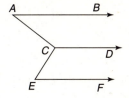

Answers: 1. They must be equal. 2. They are parallel. 3. They are parallel. 4. 104°

Chapter 3

TRIANGLES

Chapter Check-In

- ❏ Computing the measure of the third angle of a triangle, given the measures of the other two angles
- ❏ Identifying equilateral, isosceles, and scalene triangles
- ❏ Identifying the hypotenuse of a right triangle as well as altitudes, medians, and angle bisectors for any triangle
- ❏ Using the corresponding parts of two triangles to test for congruence
- ❏ Applying the Triangle Inequality Theorem

A **triangle** is a three-sided figure with three angles in its interior. The symbol for triangle is △. A triangle is named by the three letters at its vertices (the plural of vertex), a fancy name for corners. This is △*ABC* (Figure 3-1).

Figure 3-1 A triangle.

As you can imagine, the measuring of triangles and more complex figures became important long ago because of their role in surveying. Modern science has continued to find more and more practical applications requiring knowledge of triangles.

Note that any closed figure in the plane with three or more sides can be subdivided into triangles (see Figure 3-2). Consequently, what you learn about triangles can also be useful in studying more complex figures.

Figure 3-2   Triangulation of a closed figure with 5 sides.

Angle Sum of a Triangle

With the use of the *Parallel Postulate* (see Chapter 2), the following theorem can be proven.

Theorem 25: The sum of the interior angles of any triangle is 180°.

In Figure 3-1, $m\angle A + m\angle B + m\angle C = 180°$.

Example 1: If $m\angle A = 40°$ and $m\angle B = 60°$, find $m\angle C$.

Because $m\angle A + m\angle B + m\angle C = 180°$

Then, $m\angle C = 180° - (m\angle A + m\angle B)$

$m\angle C = 180° - (40° + 60°)$

$m\angle C = 80°$

Exterior Angle of a Triangle

An **exterior angle of a triangle** is formed when one side of a triangle is extended. The nonstraight angle (the one that is not just the extension of the side) outside the triangle, but adjacent to an interior angle, is an exterior angle of the triangle (Figure 3-3).

Figure 3-3   Exterior angle of a triangle.

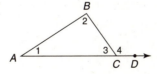

In Figure 3-3, $\angle BCD$ is an exterior angle of $\triangle ABC$.

Because $m\angle 1 + m\angle 2 + m\angle 3 = 180°$, and $m\angle 3 + m\angle 4 = 180°$, you can prove that $m\angle 4 = m\angle 1 + m\angle 2$. This is stated as a theorem.

Theorem 26: An exterior angle of a triangle is equal to the sum of the two **remote** (nonadjacent) interior angles.

Example 2: In Figure 3-3, if $m \angle 1 = 30°$ and $m \angle 2 = 100°$, find $m \angle 4$.

Because $\angle 4$ is an exterior angle of the triangle,

$$m \angle 4 = m \angle 1 + m \angle 2$$
$$m \angle 4 = 30° + 100°$$
$$m \angle 4 = 130°$$

Classifying Triangles by Sides or Angles

Triangles can be classified either according to their sides or according to their angles. All of each may be of different or the same sizes; any two sides or angles may be of the same size; there may be one distinctive angle.

The types of triangles classified by their *sides* are the following:

■ **Equilateral triangle:** A triangle with all three sides equal in measure. In Figure 3-4, the slash marks indicate equal measure.

Figure 3-4   Equilateral triangle.

■ **Isosceles triangle:** A triangle in which at least two sides have equal measure (Figure 3-5).

Figure 3-5   Isosceles triangles.

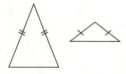

■ **Scalene triangle:** A triangle with all three sides of different measures (Figure 3-6).

Figure 3-6   Scalene triangle.

The types of triangles classified by their *angles* includes the following:

■ **Right triangle:** A triangle that has a right angle in its interior (Figure 3-7).

Figure 3-7   Right triangle.

right triangle

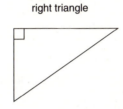

■ **Obtuse triangle:** A triangle having an obtuse angle (greater than 90° but less than 180°) in its interior. Figure 3-8 shows an obtuse triangle.

Figure 3-8   Obtuse triangle.

obtuse triangle

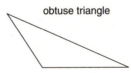

■ **Acute triangle:** A triangle having all acute angles (less than 90°) in its interior (Figure 3-9).

Figure 3-9   Acute triangle.

acute triangle

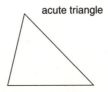

- **Equiangular triangle:** A triangle having all angles of equal measure (Figure 3-10).

Figure 3-10 Equiangular triangle.

equiangular triangle

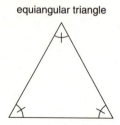

Because the sum of all the angles of a triangle is 180°, the following theorem is easily shown.

Theorem 27: Each angle of an equiangular triangle has a measure of 60°.

Special Names for Sides and Angles

Legs, base, vertex angle, and base angles. In an isosceles triangle, the two equal sides are called **legs,** and the third side is called the **base.** The angle formed by the two equal sides is called the **vertex angle.** The other two angles are called **base angles** (Figure 3-11).

Figure 3-11 Parts of an isosceles triangle.

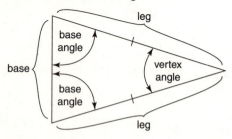

In a right triangle, the side opposite the right angle is called the **hypotenuse,** and the other two sides are called **legs** (Figure 3-12).

Figure 3-12 Parts of a right triangle.

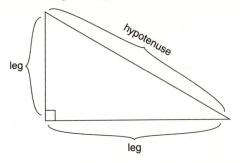

Altitudes, Medians, and Angle Bisectors

Just as there are special names for special types of triangles, so there are special names for special line segments within triangles. Now isn't that kind of special?

Base and altitude

Every triangle has three **bases** (any of its sides) and three **altitudes** (heights). Every altitude is the perpendicular segment from a vertex to its opposite side (or the extension of the opposite side) (Figure 3-13).

Figure 3-13 Three bases and three altitudes for the same triangle.

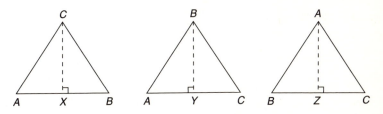

Altitudes can sometimes coincide with a side of the triangle or can sometimes meet an extended base outside the triangle. In Figure 3-14, $\overline{AC}$ is an altitude to base $\overline{BC}$, and $\overline{BC}$ is an altitude to base $\overline{AC}$.

Figure 3-14   In a right triangle, each leg can serve as an altitude.

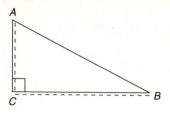

In Figure 3-15, $\overline{AM}$ is the altitude to base $\overline{BC}$

Figure 3-15   An altitude for an obtuse triangle.

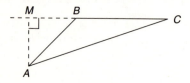

It is interesting to note that in any triangle, the three lines containing the altitudes meet in one point (Figure 3-16).

Figure 3-16   The three lines containing the altitudes intersect in a single point, which may or may not be inside the triangle.

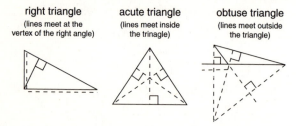

right triangle
(lines meet at the vertex of the right angle)

acute triangle
(lines meet inside the trinagle)

obtuse triangle
(lines meet outside the triangle)

Median

A **median** in a triangle is the line segment drawn from a vertex to the midpoint of its opposite side. Every triangle has three medians. In Figure 3-17, E is the midpoint of $\overline{BC}$. Therefore, $BE = EC$. $\overline{AE}$ is a median of $\triangle ABC$.

Figure 3-17 A median of a triangle.

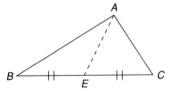

In every triangle, the three medians meet in one point inside the triangle (Figure 3-18).

Figure 3-18 The three medians meet in a single point inside the triangle.

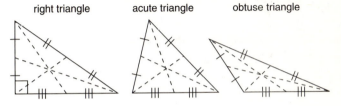

right triangle   acute triangle   obtuse triangle

Angle bisector

An **angle bisector** in a triangle is a segment drawn from a vertex that bisects (cuts in half) that vertex angle. Every triangle has three angle bisectors. In Figure 3-19, $\overline{BX}$ is an angle bisector in $\triangle ABC$.

Figure 3-19 An angle bisector.

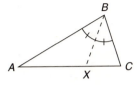

In every triangle, the three angle bisectors meet in one point inside the triangle (Figure 3-20).

Figure 3-20 The three angle bisectors meet in a single point inside the triangle.

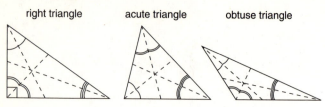

right triangle   acute triangle   obtuse triangle

In general, altitudes, medians, and angle bisectors are different segments. In certain triangles, though, they can be the same segments. In Figure 3-21, the altitude drawn from the vertex angle of an isosceles triangle can be proven to be a median as well as an angle bisector.

Figure 3-21 The altitude drawn from the vertex angle of an isosceles triangle.

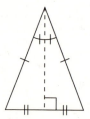

Example 3: Based on the markings in Figure 3-22, name an altitude of △*QRS;* name a median of △*QRS;* and name an angle bisector of △*QRS.*

Figure 3-22 Finding an altitude, a median, and an angle bisector.

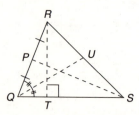

$\overline{RT}$ is an altitude to base $\overline{QS}$ because $\overline{RT} \perp \overline{QS}$.

$\overline{SP}$ is a median to base $\overline{QR}$ because P is the midpoint of $\overline{QR}$.

$\overline{QU}$ is an angle bisector of $\triangle QRS$ because it bisects $\angle RQS$.

Congruent Triangles

Triangles that have exactly the same size and shape are called **congruent triangles.** The symbol for congruent is $\cong$. Two triangles are congruent when the three sides and the three angles of one triangle have the same measurements as three sides and three angles of another triangle. The triangles in Figure 3-23 are congruent triangles.

Figure 3-23 Congruent triangles.

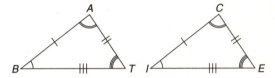

Corresponding parts

The parts of the two triangles that have the same measurements (congruent) are referred to as **corresponding parts.** This means that *Corresponding Parts of Congruent Triangles are Congruent* (CPCTC). Congruent triangles are named by listing their vertices in corresponding orders. In Figure 3-23, $\triangle BAT \cong \triangle ICE$.

Example 4: If $\triangle PQR \cong \triangle STU$, which parts must have equal measurements?

$$m \angle P = m \angle S$$
$$m \angle Q = m \angle T$$
$$m \angle R = m \angle U$$
$$PQ = ST$$
$$QR = TU$$
$$PR = SU$$

These parts are equal because corresponding parts of congruent triangles are congruent.

Tests for congruence

To show that two triangles are congruent, it is not necessary to show that all six pairs of corresponding parts are equal. The following postulates and theorems are the most common methods for proving that triangles are congruent (or equal).

Postulate 13 (SSS Postulate): If each side of one triangle is congruent to the corresponding side of another triangle, then the triangles are congruent (Figure 3-24).

Figure 3-24 The corresponding sides *(SSS)* of the two triangles are all congruent.

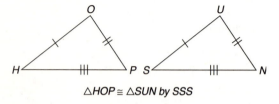

△HOP ≅ △SUN by SSS

Postulate 14 (SAS Postulate): If two sides and the angle between them in one triangle are congruent to the corresponding parts in another triangle, then the triangles are congruent (Figure 3-25).

Figure 3-25 Two sides and the included angle *(SAS)* of one triangle are congruent to the corresponding parts of the other triangle.

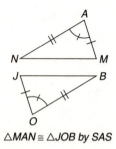

△MAN ≅ △JOB by SAS

Postulate 15 (ASA Postulate): If two angles and the side between them in one triangle are congruent to the corresponding parts in another triangle, then the triangles are congruent (Figure 3-26).

Figure 3-26 Two angles and their common side *(ASA)* in one triangle are congruent to the corresponding parts of the other triangle.

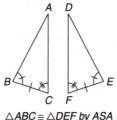

△ABC ≅ △DEF by ASA

Theorem 28 (AAS Theorem): If two angles and a side not between them in one triangle are congruent to the corresponding parts in another triangle, then the triangles are congruent (Figure 3-27).

Figure 3-27 Two angles and the side opposite one of these angles *(AAS)* in one triangle are congruent to the corresponding parts of the other triangle.

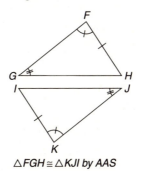

△FGH ≅ △KJI by AAS

Postulate 16 (HL Postulate): If the hypotenuse and leg of one right triangle are congruent to the corresponding parts
of another right triangle, then the triangles are
congruent (Figure 3-28).

Figure 3-28 The hypotenuse and one leg *(HL)* of the first right triangle
are congruent to the corresponding parts of the second right
triangle.

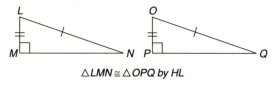

△LMN ≅ △OPQ by HL

Theorem 29 (HA Theorem): If the hypotenuse and an acute angle of one
right triangle are congruent to the corre-
sponding parts of another right triangle, then
the triangles are congruent (Figure 3-29).

Figure 3-29 The hypotenuse and an acute angle *(HA)* of the first right triangle are congruent to the corresponding parts of the sec-
ond right triangle.

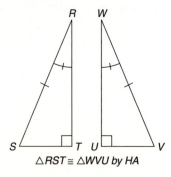

△RST ≅ △WVU by HA

Theorem 30 (LL Theorem): If the legs of one right triangle are congruent
to the corresponding parts of another right
triangle, then the triangles are congruent
(Figure 3-30).

Figure 3-30   The legs *(LL)* of the first right triangle are congruent to the corresponding parts of the second right triangle.

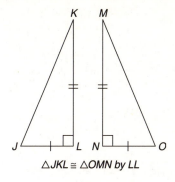

$\triangle JKL \cong \triangle OMN$ by LL

Theorem 31 (LA Theorem): If one leg and an acute angle of one right triangle are congruent to the corresponding parts of another right triangle, then the triangles are congruent (Figure 3-31).

Figure 3-31   One leg and an acute angle *(LA)* of the first right triangle are congruent to the corresponding parts of the second right triangle.

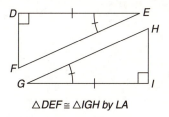

$\triangle DEF \cong \triangle IGH$ by LA

Example 5: Based on the markings in Figure 3-32, complete the congruence statement $\triangle ABC \cong \triangle \underline{}$.

$\triangle YXZ$, because A corresponds to Y, B corresponds to X, and C corresponds, to Z.

Figure 3-32   Congruent triangles.

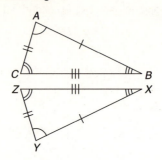

Example 6: By what method would each of the triangles in Figures 3-33(a) through 3-33(i) be proven congruent?

Figure 3-33   Methods of proving pairs of triangles congruent.

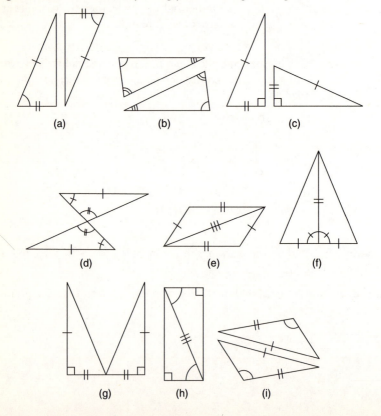

(a) *SAS.*

(b) None. There is no *AAA* method.

(c) *HL.*

(d) *AAS.*

(e) *SSS.* The third pair of congruent sides is the side that is shared by the two triangles.

(f) *SAS* or *LL.*

(g) *LL* or *SAS.*

(h) *HA* or *AAS.*

(i) None. There is no *SSA* method.

Example 7: Name the additional equal corresponding part(s) needed to prove the triangles in Figures 3-34(a) through 3-34(f) congruent by the indicated postulate or theorem.

Figure 3-34 Additional information needed to prove pairs of triangles congruent.

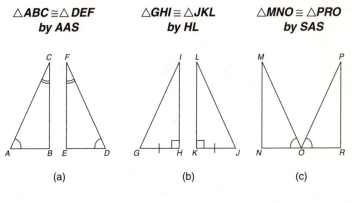

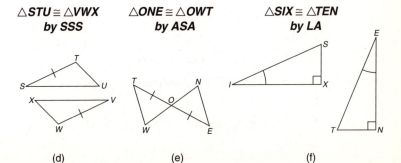

(a) *BC* = *EF* or *AB* = *DE* (*but not AC* = *DF* because these two sides lie between the equal angles).

(b) *GI* = *JL.*

(c) *MO* = *PO and NO* = RO.

(d) *TU* = *WX and SU* = *VX.*

(e) m ∠*T* = m ∠*E* and m ∠*TOW* = m ∠*EON.*

(f) *IX* = *EN* or *SX* = *TN* (*but not IS* = *ET* because they are hypotenuses).

Special Features of Isosceles Triangles

Isosceles triangles are special and because of that there are unique relationships that involve their internal line segments. Consider isosceles triangle *ABC* in Figure 3-35.

Figure 3-35 An isosceles triangle with a median.

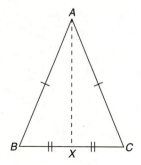

With a median drawn from the vertex to the base, $\overline{BC}$, it can be proven that *ΔBAX* ≅ *ΔCAX*, which leads to several important theorems.

Theorem 32: If two sides of a triangle are equal, then the angles opposite those sides are also equal.

Theorem 33: If a triangle is equilateral, then it is also equiangular.

Theorem 34: If two angles of a triangle are equal, then the sides opposite these angles are also equal.

Theorem 35. If a triangle is equiangular, then it is also equilateral.

Example 8: Figure 3-36 has $\triangle QRS$ with $QR = QS$. If $m\angle Q = 50°$, find $m\angle R$ and $m\angle S$.

Figure 3-36   An isosceles triangle with a specified vertex angle.

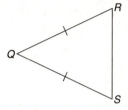

Because $m\angle Q + m\angle R + m\angle S = 180°$, and because $QR = QS$ implies that $m\angle R = m\angle S$,

$$m\angle Q + m\angle R + m\angle R = 180°$$
$$50° + 2m\angle R = 180°$$
$$2m\angle R = 130°$$
$$m\angle R = 65° \text{ and } m\angle S = 65°$$

Example 9: Figure 3-37 has $\triangle ABC$ with $m\angle A = m\angle B = m\angle C$, and $AB = 6$. Find BC and AC.

Figure 3-37   An equiangular triangle with a specified side.

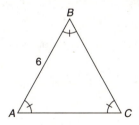

Because the triangle is equiangular, it is also equilateral. Therefore, $BC = AC = 6$.

Triangle Inequalities Regarding Sides and Angles

You have just seen that if a triangle has *equal sides,* the angles opposite these sides are equal, and if a triangle has *equal angles,* the sides opposite these angles are equal. There are two important theorems involving unequal sides and unequal angles in triangles. They are:

Theorem 36: If two sides of a triangle are unequal, then the measures of the angles opposite these sides are unequal, and the greater angle is opposite the greater side.

Theorem 37: If two angles of a triangle are unequal, then the measures of the sides opposite these angles are also unequal, and the longer side is opposite the greater angle.

Example 10: Figure 3-38 shows a triangle with angles of different measures. List the sides of this triangle in order from least to greatest.

Figure 3-38 List the sides of this triangle in increasing order.

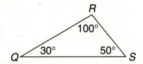

Because $30° < 50° < 100°$, then $RS < QR < QS$.

Example 11: Figure 3-39 shows a triangle with sides of different measures. List the angles of this triangle in order from least to greatest.

Figure 3-39 List the angles of this triangle in increasing order.

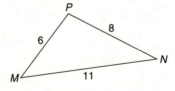

Because $6 < 8 < 11$, then $m \angle N < m \angle M < m \angle P$.

Example 12: Figure 3-40 shows right △*ABC*. Which side must be the longest?

Figure 3-40 Identify the longest side of this right triangle.

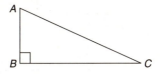

Because ∠*A* + *m* ∠ *B* + *m* ∠ *C* = 180 ° (by Theorem 25) and *m* ∠ = 90°, we have *m* ∠*A* + *m* ∠*C* = 90°. Thus, each of *m* ∠*A* and *m* ∠*C* is less than 90°. Thus ∠*B* is the angle of greatest measure in the triangle, so its opposite side is the longest. Therefore, the hypotenuse, $\overline{AC}$, is the longest side in a right triangle.

The Triangle Inequality Theorem

In △*TAB* (Figure 3-41), if *T, A,* and *B* represent three points on a map and you want to go from *T* to *B*, going from *T* to *A* to *B* would obviously be longer than going directly from *T* to *B*. The following theorem expresses this idea.

Figure 3-41 Two paths from T to B.

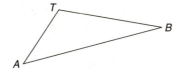

Theorem 38 (Triangle Inequality Theorem): The sum of the lengths of any two sides of a triangle is greater than the length of the third side.

Example 13: In Figure 3-42, the measures of two sides of a triangle are 7 and 12. Find the range of possibilities for the third side.

Figure 3-42   What values of x will make a triangle possible?

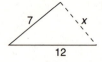

Using the *Triangle Inequality Theorem,* you can write the following:

$7 + x > 12$, so $x > 5$

$7 + 12 > x$, so $19 > x$ (or $x < 19$)

Therefore, the third side must be more than 5 and less than 19.

Chapter Checkout

Q&A

1. If a triangle has two angles with degree measure 65° and 75°, what is the degree measure of the third angle?

2. True or False: In a triangle, an angle bisector must also bisect the side opposite the angle that was bisected.

3. True or False: If each angle of a triangle is congruent to the corresponding angle of another triangle, then the two triangles must be congruent.

4. The lengths of two sides of a triangle are 11 and 23. If the third side is x, find the range of possible values for x.

Answers: 1. 40° 2. False 3. False 4. $12 < x < 34$

Chapter 4
POLYGONS

Closed shapes or figures in a plane with three or more sides are called **polygons.** Alternatively, a polygon can be defined as a closed planar figure that is the union of a finite number of line segments. In this definition, you consider *closed* as an undefined term. The term polygon is derived from a Greek word meaning "many-angled."

Some of the same topics you learned about for triangles—interior angle sum, exterior angle sum, and median, for example—will now be extended to other polygons. Just as you studied special types of triangles, you will also study special types of quadrilaterals (four-sided polygons). The chapter closes with an analogy between the median theorem of trapezoids (special quadrilaterals) and the midpoint theorem of triangles.

Classifying Polygons

Polygons first fit into two general categories—**convex** and **not convex** (sometimes called **concave**). Figure 4-1 shows some convex polygons, some non-convex polygons, and some figures that are not even classified as polygons. See the Glossary for the definitions of convex and concave polygons.

Figure 4-1 Which are polygons? Which of the polygons are convex?

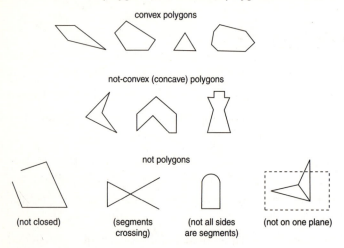

Identifying the parts of a polygon

The endpoints of the sides of polygons are called **vertices.** When naming a polygon, its vertices are named in consecutive order either clockwise or counterclockwise.

Consecutive sides are two sides that have an endpoint in common. The four-sided polygon in Figure 4-2 could have been named *ABCD, BCDA,* or *ADCB,* for example. It does not matter with which letter you begin as long as the vertices are named consecutively. Sides $\overline{AB}$ and $\overline{BC}$ are examples of consecutive sides.

Figure 4-2 There are four pairs of consecutive sides in this polygon.

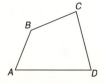

A **diagonal** of a polygon is any segment that joins two nonconsecutive vertices. Figure 4-3 shows five-sided polygon *QRSTU*. Segments $\overline{QS}$, $\overline{SU}$, $\overline{UR}$, $\overline{RT}$, and $\overline{QT}$ are the diagonals in this polygon.

Figure 4-3 Diagonals of a polygon.

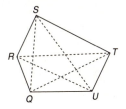

Number of sides

Polygons are also classified by how many sides (or angles) they have. The following lists the different types of polygons and the number of sides that they have:

A **triangle** is a three-sided polygon.

A **quadrilateral** is a four-sided polygon.

A **pentagon** is a five-sided polygon.

A **hexagon** is a six-sided polygon.

A **septagon** or heptagon is a seven-sided polygon.

An **octagon** is an eight-sided polygon.

A **nonagon** is a nine-sided polygon.

A **decagon** is a ten-sided polygon.

It was shown earlier that an equilateral triangle is automatically equiangular and that an equiangular triangle is automatically equilateral. This does not hold true for polygons in general, however. Figure 4-4 shows examples of quadrilaterals that are equiangular but not equilateral, equilateral but not equiangular, and equiangular and equilateral.

Figure 4-4 An equiangular quadrilateral does not have to be equilateral, and an equilateral quadrilateral does not have to be equiangular.

equiangular but   equilateral but   equiangular and
not equilateral   not equiangular   equilateral

Regular polygons

When a polygon is both equilateral and equiangular, it is referred to as a **regular polygon.** For a polygon to be regular, it must also be convex. Figure 4-5 shows examples of regular polygons.

Figure 4-5 Regular polygons.

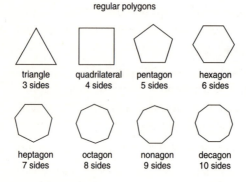

regular polygons

triangle   quadrilateral   pentagon   hexagon
3 sides   4 sides   5 sides   6 sides

heptagon   octagon   nonagon   decagon
7 sides   8 sides   9 sides   10 sides

Angle Sum of Polygons

When you begin with a polygon with four or more sides and draw all the diagonals possible from one vertex, the polygon then is divided into several nonoverlapping triangles. Figure 4-6 illustrates this division using a seven-sided polygon. The **interior angle sum** of this polygon can now be found by multiplying the number of triangles by 180°. Upon investigating, it is found that the number of triangles is always two less than the number of sides. This fact is stated as a theorem.

Theorem 39: If a convex polygon has *n* sides, then its interior angle sum is given by the following equation: $S = (n - 2) \times 180°$.

Figure 4-6 Triangulation of a seven-sided polygon to find the interior angle sum.

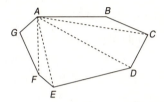

The polygon in Figure 4-6 has seven sides, so using *Theorem 39* gives:

(7 - 2) nonoverlapping triangles = 5 nonoverlapping triangles

interior angle sum = $5 \times 180°$

interior angle sum = 900°

An **exterior angle of a polygon** is formed by extending only one of its sides. The nonstraight angle adjacent to an interior angle is the exterior angle. Figure 4-7 might suggest the following theorem:

Theorem 40: If a polygon is convex, then the sum of the degree measures of the exterior angles, one at each vertex, is 360°.

$$m \angle 1 + m \angle 2 + m \angle 3 + m \angle 4 + m \angle 5 + m \angle 6 = 360°$$

Figure 4-7 The (nonstraight) exterior angles of a polygon.

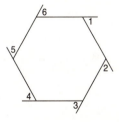

Example 1: Find the interior angle sum of a decagon.

A decagon has 10 sides, so:

$$S = (10 - 2) \times 180°$$
$$S = 1440°$$

Example 2: Find the exterior angle sums, one exterior angle at each vertex, of a convex nonagon.

The sum of the exterior angles of any convex polygon is 360°.

Example 3: Find the measure of each interior angle of a regular hexagon (Figure 4-8).

Figure 4-8 An interior angle of a regular hexagon.

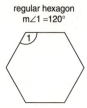

regular hexagon
$m\angle 1 = 120°$

Method 1: Because the polygon is regular, all interior angles are equal, so you only need to find the interior angle sum and divide by the number of angles.

$$S = (6 - 2) \times 180°$$

$$S = 720$$

There are six angles, so $720 \div 6 = 120°$

Each interior angle of a regular hexagon has a measure of 120°.

Method 2: Because the polygon is regular and all its interior angles are equal, all its exterior angles are also equal. Look at Figure 4-7. This means that

$$m\angle 1 = m\angle 2 = m\angle 3 = m\angle 4 = m\angle 5 = m\angle 6$$

Because the sum of these angles will always be 360°, then each exterior angle would be 60° ($360° \div 6 = 60°$). If each exterior angle is 60°, then each interior angle is 120° ($180° - 60° = 120°$).

Special Quadrilaterals

Unlike humans, all quadrilaterals are not created equal. It's not a matter of size I'm alluding to here, but rather a question of features. They may have a pair of parallel sides, two pairs, a right angle . . . but I'm getting ahead of myself.

Trapezoid

A **trapezoid** is a quadrilateral with only one pair of opposite sides parallel. The parallel sides are called **bases,** and the *non*parallel sides are called legs. A segment that joins the midpoints of the legs is called the **median of the trapezoid.** Any segment that is perpendicular to both bases is called an **altitude of the trapezoid** (Figure 4-9). The length of an altitude is called the **height** of the trapezoid.

Figure 4-9   A trapezoid with its median and an altitude.

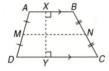

$\overline{AB}$ and $\overline{CD}$ are bases.

$\overline{XY}$ is an altitude.

$\overline{MN}$ is the median.

XY, length of segment $\overline{XY}$, is the height.

Parallelogram

A **parallelogram** is any quadrilateral with both pairs of opposite sides parallel. Each pair of parallel sides is called a pair of **bases of the parallelogram.** Any perpendicular segment between a pair of bases is called the an **altitude of the parallelogram.** The length of an altitude is the height of the parallelogram. The symbol ▱ is used for the word parallelogram. Figure 4-10 shows that a parallelogram has two sets of bases and that, with each set of bases, there is an associated height.

Figure 4-10   A parallelogram with its bases and associated heights.

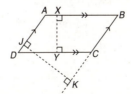

In $\square ABCD$,

$\overline{XY}$ is an altitude to bases $\overline{AB}$ and $\overline{CD}$.

$\overline{JK}$ is an altitude to bases $\overline{AD}$ and $\overline{BC}$.

XY is the height of $\square ABCD$, with $\overline{AB}$ and $\overline{CD}$ as bases, JK is the height of $\square ABCD$, with $\overline{AD}$ and $\overline{BC}$ as bases.

The following are theorems regarding parallelograms:

Theorem 41: A diagonal of a parallelogram divides it into two congruent triangles.

In $\square ABCD$ with diagonal $\overline{BD}$, according to *Theorem 41*, $\triangle ABD \cong \triangle CDB$ (Figure 4-11).

Figure 4-11 Two congruent triangles created by a diagonal of a parallelogram.

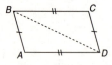

Theorem 42: Opposite sides of a parallelogram are congruent.

Theorem 43: Opposite angles of a parallelogram are congruent.

Theorem 44: Consecutive angles of a parallelogram are supplementary.

In $\square ABCD$ (Figure 4-12):

■ By *Theorem 42*, $AB = DC$ and $AD = BC$.

■ By *Theorem 43*, $m \angle A = m \angle C$ and $m \angle B = m \angle D$.

■ By *Theorem 44:*

    $\angle A$ and $\angle B$ are supplementary.

    $\angle B$ and $\angle C$ are supplementary.

    $\angle C$ and $\angle D$ are supplementary.

    $\angle A$ and $\angle D$ are supplementary.

Figure 4-12 A parallelogram.

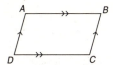

Theorem 45: The diagonals of a parallelogram bisect each other.

In ☐ *ABCD* (Figure 4-13), by *Theorem 45, AE = EC* and *BE = ED.*

Figure 4-13 The diagonals of a parallelogram bisect one another.

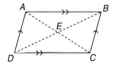

Proving That Figures Are Parallelograms

Many times you will be asked to prove that a figure is a parallelogram. The following theorems are tests that determine whether a quadrilateral is a parallelogram:

Theorem 46: If both pairs of opposite sides of a quadrilateral are equal, then it is a parallelogram.

Theorem 47: If both pairs of opposite angles of a quadrilateral are equal, then it is a parallelogram.

Theorem 48: If all pairs of consecutive angles of a quadrilateral are supplementary, then it is a parallelogram.

Theorem 49: If one pair of opposite sides of a quadrilateral is both equal and parallel, then it is a parallelogram.

Theorem 50: If the diagonals of a quadrilateral bisect each other, then it is a parallelogram.

Figure 4-14 A quadrilateral with its diagonals.

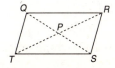

Quadrilateral *QRST* in Figure 4-14 is a parallelogram if:

- *QR = ST* and *QT = RS*, by *Theorem 46.*

- *m ∠Q = m ∠S* and *m ∠T = m ∠R*, by *Theorem 47.*

- *∠Q* and *∠R, ∠R* and *∠S, ∠S* and *∠T*, and *∠Q* and *∠T* are all supplementary pairs, by *Theorem 48.*

- *QR = ST* and $\overline{QR} /\!/ \overline{ST}$ or *QT = RS* and $\overline{QT} /\!/ \overline{RS}$, by *Theorem 49.*

- *QP = PS* and *RP = PT*, by *Theorem 50.*

Properties of Special Parallelograms

If it is true that not all quadrilaterals are created equal, the same may be said about parallelograms. You can even out the sides or stick in a right angle.

Rectangle

A **rectangle** is a quadrilateral with all right angles. It is easily shown that it must also be a parallelogram, with all of the associated properties. A rectangle has an additional property, however.

Theorem 51: The diagonals of a rectangle are equal.

In rectangle *ABCD* (Figure 4-15), *AC = BD,* by *Theorem 51.*

Figure 4-15 The diagonals of a rectangle are equal.

Rhombus

A **rhombus** is a quadrilateral with all equal sides. It is also a parallelogram with all of the associated properties. A rhombus, however, also has additional properties.

Theorem 52: The diagonals of a rhombus bisect opposite angles.

Theorem 53: The diagonals of a rhombus are perpendicular to one another.

In rhombus *CAND* (Figure 4-16), by *Theorem 52,* $\overline{CN}$ bisects ∠*DCA* and ∠*DNA*. Also, $\overline{AD}$ bisects ∠*CAN* and ∠*CDN* and by *Theorem 53,* $\overline{CN} \perp \overline{AD}$.

Figure 4-16   The diagonals of a rhombus are perpendicular to one another and bisect opposite angles.

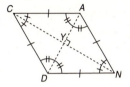

Square

A **square** is a quadrilateral with all right angles and all equal sides. A square is also a parallelogram, a rectangle, and a rhombus and has all the properties of all these special quadrilaterals. Figure 4-17 shows a square.

Figure 4-17   A square has four right angles and four equal sides.

Figure 4-18 summarizes the relationships of these quadrilaterals to one another.

Figure 4-18   The relationships among the various types of quadrilaterals.

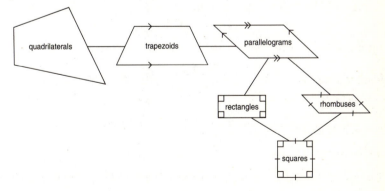

Example 4: Identify the following figures.

Figure 4-19 Identify these polygons.

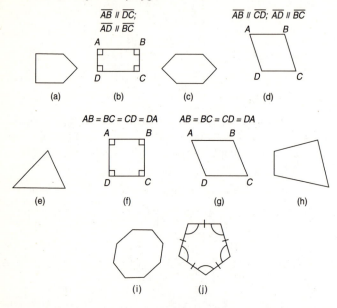

(a) pentagon, (b) rectangle, (c) hexagon, (d) parallelogram, (e) triangle, (f) square, (g) rhombus, (h) quadrilateral, (i) octagon, and (j) regular pentagon

Example 5: In Figure 4-20, find $m \angle A$, $m \angle C$, $m \angle D$, CD, and AD.

Figure 4-20 A parallelogram with one angle specified.

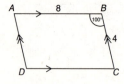

$m \angle A = m \angle C = 80°$, because consecutive angles of a parallelogram are supplementary.

$m \angle D = 100°$, because opposite angles of a parallelogram are equal.

$CD = 8$ and $AD = 4$, because opposite sides of a parallelogram are equal.

Example 6: In Figure 4-21, find *TR, QP, PS, TP,* and *PR.*

Figure 4-21 A rectangle with one diagonal specified.

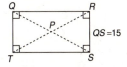

TR = 15, because diagonals of a rectangle are equal.

$QP = PS = TP = PR = 7.5$, because diagonals of a rectangle bisect each other.

Example 7: In Figure 4-22, find $m \angle MOE$, $m \angle NOE$, and $m \angle MYO$.

Figure 4-22 A rhombus with one angle specified.

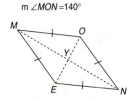

$m \angle MOE = m \angle NOE = 70°$, because diagonals of a rhombus bisect opposite angles.

$m \angle MYO = 90°$, because diagonals of a rhombus are perpendicular.

Properties of Trapezoids

Recall that a trapezoid is a quadrilateral with only one pair of opposite sides parallel and that the parallel sides are called bases and the nonparallel sides are called legs. If the legs of a trapezoid are equal, it is called **an isosceles trapezoid.** Figure 4-23 is an isosceles trapezoid.

Figure 4-23 An isosceles trapezoid.

A pair of angles that share the same base are called **base angles** of the trapezoid. In Figure 4-23, ∠A and ∠B or ∠C and ∠D are base angles of trapezoid *ABCD*. Two special properties of an isosceles trapezoid can be proven.

Theorem 53: Base angles of an isosceles trapezoid are equal.

Theorem 54: Diagonals of an isosceles trapezoid are equal.

In isosceles trapezoid *ABCD* (Figure 4-24) with bases $\overline{AB}$ and $\overline{CD}$:

■ By *Theorem 53*, $m \angle DAB = m \angle CBA$, and $m \angle ADC = m \angle BCD$.

■ By *Theorem 54*, $AC = BD$.

Figure 4-24 An isosceles trapezoid with its diagonals.

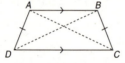

Recall that the median of a trapezoid is a segment that joins the midpoints of the nonparallel sides.

Theorem 55: The median of any trapezoid has two properties: (1) It is parallel to both bases. (2) Its length equals half the sum of the base lengths.

In trapezoid *ABCD* (Figure 4-25) with bases $\overline{AB}$ and $\overline{CD}$, *E* the midpoint of $\overline{AD}$, and *F* the midpoint of $\overline{BC}$, by *Theorem 55*:

$$\overline{EF} \parallel \overline{AB}$$

$$\overline{EF} \parallel \overline{CD}$$

$$EF = 1/2\,(AB + CD)$$

Figure 4-25 A trapezoid with its median.

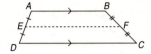

Example 8: In Figure 4-26, find $m \angle ABC$ and find BD.

Figure 4-26 An isosceles trapezoid with a specified angle and a specified diagonal.

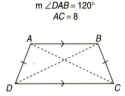

$m \angle ABC = 120°$, because the base angles of an isosceles trapezoid are equal.

$BD = 8$, because diagonals of an isosceles trapezoid are equal.

Example 9: In Figure 4-27, find TU.

Figure 4-27 A trapezoid with its two bases given and the median to be computed.

$QR = 15$ and $PS = 25$

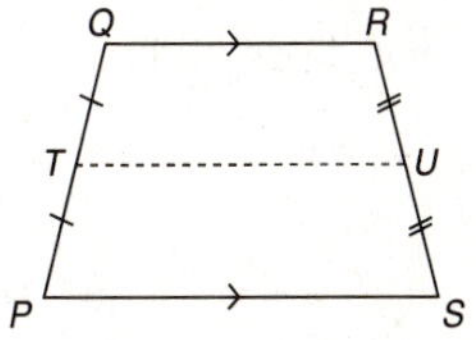

Because the median of a trapezoid is half the sum of the lengths of the bases:

$$TU = 1/2(15 + 25)$$
$$TU = 1/2 \ (40)$$
$$TU = 20$$

The Midpoint Theorem

Figure 4-28 shows △*ABC* with *D* and *E* as midpoints of sides $\overline{AC}$ and $\overline{AB}$ respectively. If you look at this triangle as though it were a trapezoid with one base of $\overline{BC}$ and the other base so small that its length is virtually zero, you could apply the "median" theorem of trapezoids, *Theorem 55.*

Theorem 56 (Midpoint Theorem): The segment joining the midpoints of two sides of a triangle is parallel to the third side and half as long as the third side.

In Figure 4-28, by *Theorem 56,*

$$\overline{DE} \,/\!/\, \overline{BC}$$

$$DE = (1/2)\ (BC)$$

Figure 4-28   The segment joining the midpoints of two sides of a triangle.

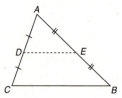

Example 10: In Figure 4-29, find *HJ.*

Figure 4-29   Compute the length of the broken line segment joining the midpoints of two sides of the triangle.

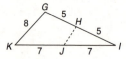

Because *H* and *J* are midpoints of two sides of a triangle:

$$HJ = 1/2\ GK$$
$$HJ = 1/2(8)$$
$$HJ = 4$$

Chapter Checkout

Q&A

1. True or False: All triangles are convex.

2. What is the degree measure of the interior angle determined by two adjacent sides of a regular decagon?

Answers: 1. True 2. 144°

Chapter 5

PERIMETER AND AREA

Chapter Check-In

❑ Computing the perimeter and the area of a square, a rectangle, a parallelogram, a triangle, a trapezoid, and a regular polygon

❑ Computing the circumference and the area of a circle

Perimeter refers to the entire length of a figure or the distance around it. If the figure is a circle, the length is referred to as the **circumference.** Such lengths are always measured in linear units such as inches, feet, and centimeters. **Area** refers to the size of the interior of a planar (flat) figure. Area is always measured in square units such as square inches (in^2), square feet (ft^2), and square centimeters (cm^2), or in special units such as acres or hectares.

Squares and Rectangles

For all polygons, you find perimeter by adding together the lengths of all the sides. In this section, *P* is used to stand for *perimeter,* and *A* is used to stand for *area.*

Finding the perimeter

Figures 5-1(a) and 5-1(b) show perimeter formulas for squares and rectangles.

Figure 5-1 Perimeter of a square and perimeter of a rectangle.

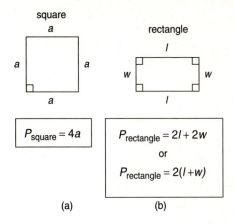

(a)   (b)

Finding the area

Area formulas for squares and rectangles are formed by simply multiplying any pair of consecutive sides together. Refer to Figures 5-1(a) and 5-1(b).

$$A_{\text{square}} = a^2 \qquad A_{\text{rectangle}} = lw$$

Example 1: Find the perimeter and area of Figure 5-2.

Figure 5-2 Finding the perimeter and area of a square.

8 in

This is a square.

$$P_{\text{square}} = 4a \qquad\qquad A_{\text{square}} = a^2$$
$$= 4(8 \text{ in}) \qquad\qquad = (8 \text{ in})^2$$
$$= 32 \text{ in} \qquad\qquad = 64 \text{ in}^2$$

Example 2: Find the perimeter and area of Figure 5-3.

Figure 5-3 Finding the perimeter and area of a rectangle.

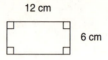

This is a rectangle.

$$P_{\text{rectangle}} = 2(l + w) \qquad A_{\text{rectangle}} = lw$$
$$= 2(12 \text{ cm} + 6 \text{ cm}) \qquad = (12 \text{ cm})(6 \text{ cm})$$
$$= 36 \text{ cm} \qquad = 72 \text{ cm}^2$$

Example 3: If the perimeter of a square is 36 ft, find its area.

$$P_{\text{square}} = 4a \qquad A_{\text{square}} = a^2$$
$$36 \text{ ft} = 4a \qquad = (9 \text{ ft})^2$$
$$9 \text{ ft} = a \qquad = 81 \text{ ft}^2$$

The area of the square would be 81 square feet.

Example 4: If a rectangle with length 9 in has an area of 36 in², find its perimeter.

$$A_{\text{rectangle}} = lw \qquad P_{\text{square}} = 2(l + w)$$
$$36 \text{ in}^2 = (9 \text{ in})(w) \qquad = 2(9 \text{ in} + 4 \text{ in})$$
$$4 \text{ in} = w \qquad = 26 \text{ in}$$

The perimeter of the rectangle would be 26 inches.

Parallelograms

In the parallelogram shown in Figure 5-4, *h* is a height because it is perpendicular to a pair of opposite sides called bases. One of the bases has been labeled *b*, and the nonbase remaining sides are each labeled *a*.

Figure 5-4 A parallelogram with base and height labeled.

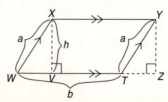

Finding the perimeter

The following formula is now apparent.

$$P_{parallelogram} = 2(a + b) \text{ or } P_{parallelogram} = 2a + 2b$$

Finding the Area

In Figure 5-4, also notice that $\triangle WXV \cong \triangle TYZ$, which means that they also have equal areas. This makes the area of $\square\,WXYT$ the same as the area of $\square\,XYZV$. But $A_{rectangle}\,XYZV = bh$, so $A_{parallelogram}\,XYTW = bh$. That is, the area of a parallelogram is the product of any base with its respective height.

$$A_{parallelogram} = bh$$

Example 5: Find the perimeter and area of Figure 5-5.

Figure 5-5 Finding the perimeter and area of a parallelogram.

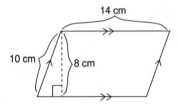

The figure is a parallelogram, so

$$P_{parallelogram} = 2(a + b) \qquad\qquad A_{parallelogram} = bh$$
$$= 2(10 \text{ cm} + 14 \text{ cm}) \qquad\qquad = (14 \text{ cm})(8 \text{ cm})$$
$$= 48 \text{ cm} \qquad\qquad\qquad\qquad = 112 \text{ cm}^2$$

Triangles

Look at $\triangle ABD$ in Figure 5-6. If a line $\overline{BC}$ is drawn through B parallel to $\overline{AD}$ and another line $\overline{DC}$ is drawn through D parallel to $\overline{AB}$, then you will have formed a parallelogram. $\overline{BD}$ is now a diagonal in this parallelogram. Because a diagonal divides a parallelogram into two congruent triangles, the area of $\triangle ABD$ is exactly half the area of $\square\,ABCD$.

Figure 5-6   Area of a triangle is half the area of the associated parallelogram.

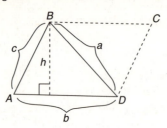

Finding the Area

Because $A_{\text{parallelogram}} = bh$, then

$$A_{\text{triangle}} = 1/2\,bh$$

Finding the Perimeter

In $\triangle ABD$ (Figure 5-6), the perimeter is found simply by adding the lengths of the three sides.

$$P_{\text{triangle}} = a + b + c$$

Example 6: Find the perimeter and area for the triangles in Figures 5-7(a), 5-7(b), and 5-7(c).

Figure 5-7   Finding perimeters and areas of triangles.

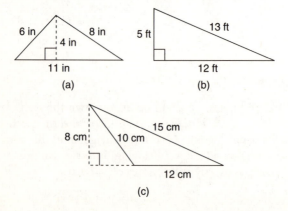

(a) $P_{triangle} = a + b + c$
$\qquad = 8 + 11 + 6$
$\qquad = 25$ in

$A_{triangle} = 1/2bh$
$\qquad = 1/2(11)(4)$
$\qquad = 22$ in^2

(b) $P_{triangle} = a + b + c$
$\qquad = 13 + 12 + 5$
$\qquad = 30$ ft

$A_{triangle} = 1/2bh$
$\qquad = 1/2(12)(5)$
$\qquad = 30$ ft^2

(c) $P_{triangle} = a + b + c$
$\qquad = 15 + 12 + 10$
$\qquad = 37$ cm

$A_{triangle} = 1/2bh$
$\qquad = 1/2(12)(8)$
$\qquad = 48$ cm^2

Example 7: If the area of a triangle is 64 cm^2 and it has a height of 16 cm, find the length of its base.

$$A_{triangle} = 1/2bh$$
$$64 \text{ cm}^2 = 1/2(b)(16 \text{ cm})$$

Multiply both sides by 2.

$$128 \text{ cm}^2 = (b)(16 \text{ cm})$$
$$8 \text{ cm} = b$$

The triangle will have a base of 8 centimeters.

Trapezoids

Probably the trapezoid is one of the most popular quadrilaterals when it comes to bridge construction. Numerous railroad trestles and wooden bridges of the nineteenth and early twentieth centuries were trapezoidal in shape.

Finding the Perimeter

In Figure 5-8, trapezoid *QRSV* is labeled so that b_1 and b_2 are the bases (*h* is the height to these bases) and *a* and *c* are the legs. The perimeter is simply the sum of these lengths.

Figure 5-8   A trapezoid and the associated parallelogram.

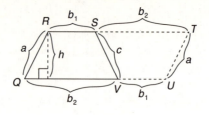

$$P_{\text{trapezoid}} = a + b_1 + c + b_2$$

Finding the Area

Referring to Figure 5-8, an identical, but upside-down trapezoid is drawn adjacent to trapezoid *QRSV*, trapezoid *TUVS*. It can now be shown that the figure *QRTU* is a parallelogram, and its area can now be found.

$$A_{\text{parallelogram}} \; QRTU = (\text{base})(\text{height})$$
$$= (b_1 + b_2)h$$

Because trapezoid *QRSV* is exactly half of this parallelogram, the following formula gives the area of a trapezoid.

$$A_{\text{trapezoid}} = 1/2(b_1 + b_2)h$$

Example 8: Find the perimeter and area of Figure 5-9.

Figure 5-9   Finding the perimeter and area of a trapezoid.

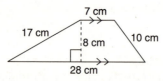

The figure is a trapezoid.

$$P_{\text{trapezoid}} = a + b_1 + c + b_2 \qquad A_{\text{trapezoid}} = 1/2(b_1 + b_2)h$$
$$= 17 + 7 + 10 + 28 \qquad\qquad = 1/2(7 + 28)(8)$$
$$= 62 \text{ cm} \qquad\qquad\qquad = 140 \text{ cm}^2$$

Regular Polygons

Thus far, we have dealt with polygons of three and four sides. But there is really no limit to the number of sides a polygon may have. The only practical limit is that unless you draw them on a very large sheet of paper, after about 20 sides or so, the polygon begins to look very much like a circle.

Parts of a regular polygon

In a regular polygon, there is one point in its interior that is equidistant from its vertices. This point is called the **center** of the **regular polygon.** In Figure 5-10, O is the center of the regular polygon.

The **radius** of a regular polygon is a segment that goes from the center to any vertex of the regular polygon.

The **apothem** of a regular polygon is any segment that goes from the center and is perpendicular to one of the polygon's sides. In Figure 5-10, $\overline{OC}$ is a radius and $\overline{OX}$ is an apothem.

Figure 5-10   Center, radius, and apothem of a regular polygon.

Finding the Perimeter

Because a regular polygon is equilateral, to find its perimeter you need to know only the length of one of its sides and multiply that by the number of sides. Using n-gon to represent a polygon with n sides, and s as the length of each side, produces the following formula.

$$P_{\text{regular } n\text{-gon}} = ns$$

Finding the Area

If p represents the perimeter of the regular polygon and a represents the length of its apothem, the following formula can eventually be shown to represent its area.

$$A_{\text{regular } n\text{-gon}} = 1/2\,ap$$

Example 9: Find the perimeter and area of the regular pentagon in Figure 5-11, with apothem approximately 5.5 in.

Figure 5-11 Finding the perimeter and area of a regular pentagon.

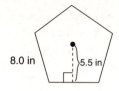

8.0 in   5.5 in

$P_{\text{regular } n\text{-gon}} = ns$ $A_{\text{regular } n\text{-gon}} = 1/2 ap$
$= (5)(8.0)$ $= 1/2(5.5)(40)$
$= 40 \text{ in}$ $= 110 \text{ in}^2$

Circles

A **circle** is a planar figure with all points the same distance from a fixed point. That fixed point is called the **center of the circle.** Any segment that goes from the center to a point on the circle is called a **radius of the circle.** A **diameter** is any segment that passes through the center and has its endpoints on the circle. Obviously, a diameter is twice as long as a radius. In Figure 5-12, O is the center, $\overline{OB}$, $\overline{OC}$, and $\overline{OA}$ each a radius, and $\overline{AC}$ is a diameter.

Figure 5-12 A circle with center, radius, and diameter labeled.

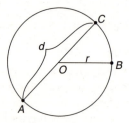

Finding the Circumference

In ancient times the Greeks discovered that for all circles, the circumference divided by the diameter always turns out to be the same constant value. The Greek letter π (pi) is now used to represent that value. In fractional or decimal form, the commonly used approximations of π are $\pi \approx 3.14$ or $\pi \approx 22/7$. The Greeks found the formula $C_{\text{circle}}/d = \pi$, which is rewritten in the following form.

$$C_{\text{circle}} = \pi d \text{ or } C_{\text{circle}} = 2\pi r$$

If you briefly regard a circle as a regular polygon with infinitely many infinitesimally small sides, you see that the apothem and radius become the same length (Figure 5-13).

Figure 5-13 Apothem and radius of a circle.

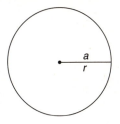

Finding the Area

Looking at the area formula for a regular polygon and making the appropriate changes with regard to the circle,

$$A_{\text{regular } n\text{-gon}} = 1/2(a)(p)$$

$$A_{\text{circle}} = 1/2(r)(2\pi r)$$

$$= 1/2(2\pi r)(r)$$

$$= \pi r^2$$

That is, the formula for the area of a circle now becomes the following:

$$A_{\text{circle}} = \pi r^2$$

Example 10: Find the circumference and area for the circle in Figure 5-14. Use 3.14 as an approximation for π.

Figure 5-14 Finding the circumference and area of a circle.

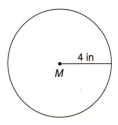

4 in

M

$$C_{circle} = \pi d$$
$$= (3.14)(8)$$
$$= 25.12 \text{ in}$$

$$A_{circle} = \pi r^2$$
$$= (3.14)(4)^2$$
$$= 50.24 \text{ in}^2$$

Example 11: If the area of a circle is 81π ft^2, find its circumference.

$$A_{circle} = \pi r^2$$
$$81\pi \text{ ft}^2 = \pi r^2$$
$$9 \text{ ft} = r$$
$$d = 18 \text{ ft}$$

$$C_{circle} = \pi d$$
$$= \pi(18)$$
$$= 18\pi \text{ ft}$$

So the circumference is approximately 56.52 ft.

Summary of Perimeter, Circumference, and Area Formulas

With a simple formula, you can find the perimeter or area for any shape. Figure 5-15 is a summary of formulas for each shape.

Figure 5-15: Summary of Perimeter and Area Formulas.

Figure	Name	Perimeter/Circumference	Area
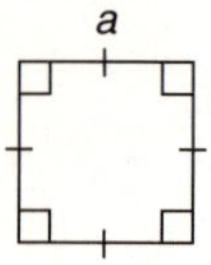 (a)	square	$4a$	a^2
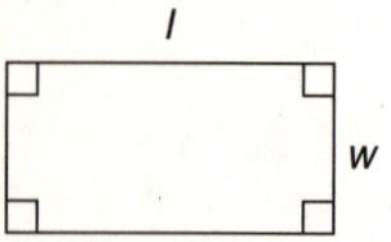 (b)	rectangle	$2l + 2w$ or $2(l+w)$	lw
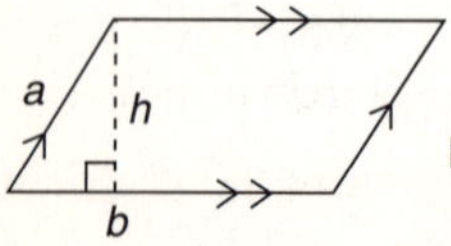 (c)	parallelogram	$2a + 2b$ or $2(a+b)$	bh

Figure	Name	Perimeter/ Circumference	Area
(d)	triangle	$a + b + c$	$1/2bh$
(e)	trapezoid	$a + b_1 + c + b_2$	$1/2(b_1+b_2)h$
(f)	regular polygon	ns n = number of sides	$1/2ap$ p = perimeter a = apothem
(g)	circle	πd or $2\pi r$	πr^2

Chapter Checkout

Q&A

1. Compute the perimeter and the area of a rectangle with base 7 cm and height 4 cm.

2. Compute the perimeter and the area of an isosceles triangle with height 12 in and whose sides are 10 in, 13 in, and 13 in.

3. Compute the circumference and the area of a circle of radius 5 in.

Answers: 1. $P = 22$ cm, $A = 28$ cm^2 2. $P = 36$ in, $A = 60$ in^2 3. $C = 10\pi \approx 31.4$ in, $A = 25\pi \approx 78.5$ in^2

Chapter 6
SIMILARITY

Chapter Check-In

❏ Applying the *Cross-Product* (or *Means-Extremes*) *Property* of proportions

❏ Knowing what is required for two polygons to be similar and applying proportions to similar polygons (including triangles)

❏ Applying the *AA Similarity Postulate,* the *Side-Splitter Theorem,* and the *Angle Bisector Theorem* for triangles

❏ Applying the scale factor to similar triangles

As you have learned in Chapters 4 and 5, two polygons that have exactly the same size and shape are *congruent.* This chapter introduces similarity, a less restrictive property than congruence. For two polygons to be **similar**, they must have the same shape but not necessarily the same size. The study of similar triangles led to the definitions of the six trigonometric functions (not presented here) and thus to the branch of mathematics known as trigonometry.

Ratio and Proportion

Ratio is a concept that you have probably encountered in other math classes. It is a comparison of sizes.

Ratio

The **ratio** of two numbers a and b is the fraction $\frac{a}{b}$, usually expressed in reduced form. An alternative form involves a colon. The colon form is most frequently used when comparing three or more numbers to each other. See Table 6-1.

Table 6-1   Ratio Formats

Ratio	Written Form
3 to 4	3/4 or 3 : 4
a to b, $b \neq 0$	a/b or $a : b$
1 to 3 to 5	1 : 3 : 5

Example 1: A classroom has 25 boys and 15 girls. What is the ratio of boys to girls?

> boys to girls = 25 to 15 or 25 : 15, which reduces to
> > = 5 to 3 or 5 : 3

The ratio of boys to girls is 5 to 3, or 5/3, or 5 : 3.

Example 2: The ratio of two supplementary angles is 2 to 3. Find the measure of each angle.

> Let measure of smaller angle = $2x$, measure of larger angle = $3x$.
> $2x$ to $3x$ reduces to 2 to 3.
> $2x + 3x = 180°$ (The sum of supplementary angles is 180°.)
> $5x = 180°$
> $x = 36°$
> Then, $2x = 2(36°)$ and $3x = 3(36°)$.
> So, $2x = 72°$ and $3x = 108°$

The angles have measures of 72° and 108°.

Example 3: A triangle has angle measures of 40°, 50°, and 90°. In simplest form, what is the ratio of these angles to each other?

> 40 : 50 : 90 = 4 : 5 : 9 (10 is a common divisor)

This means that:

> (1) The ratio of the first to the second is 4 to 5.
> (2) The ratio of the first to the third is 4 to 9.
> (3) The ratio of the second to the third is 5 to 9.

Example 4: A 50-inch segment is divided into three parts whose lengths have the ratio 2 : 3 : 5. What is the length of the longest part?

Let measure of shortest piece = $2x$

measure of middle piece = $3x$

measure of longest piece = $5x$

$$2x + 3x + 5x = 50$$

$$10x = 50$$

$$x = 5$$

$2x = 2(5)$	$3x = 3(5)$	$5x = 5(5)$
$2x = 10$	$3x = 15$	$5x = 25$

The longest part has a measure of 25 inches.

Proportion

A **proportion** is an equation stating that two ratios are equal.

$$\frac{8}{10} = \frac{4}{5} \qquad 8 : 10 = 4 : 5$$

Means and extremes

The extremes are the terms in a proportion that are the farthest apart when the proportion is written in colon form ($a{:}b = c{:}d$). In the foregoing, a and d are extremes. The means are the two terms closest to each other.

In the preceding proportion, the values a and d are called extremes of the proportion; the values b and c are called the means of the proportion.

Properties of Proportions

The four properties that follow are not difficult to justify algebraically, but the details will not be presented here.

Property 1 (Means-Extremes Property, or Cross-Products Property): If $a/b = c/d$, then $ad = bc$. Conversely, if $ad = bc \neq 0$, then $\frac{a}{b} = \frac{c}{d}$ and $\frac{b}{a} = \frac{d}{c}$.

$$8/10 = 4/5 \text{ is a proportion}$$

Property 1 states $(8)(5) = (10)(4)$

$$40 = 40$$

Example 5: Find a if $a/12 = 3/4$.

By *Property 1:*

$$(a)(4) = (12)(3)$$

$$4a = 36$$

$$a = 9$$

Example 6: Is $3 : 4 = 7 : 8$ a proportion?

No. If this were a proportion, *Property 1* would produce

$$(3)(8) = (4)(7)$$

$$24 = 28, \text{ which is not true}$$

Property 2 (Means or Extremes Switching Property): If $a/b = c/d$ and is a proportion, then both $d/b = c/a$ and $a/c = b/d$ are proportions.

Example 7: $8/10 = 4/5$ is a proportion. *Property 2* says that if you were to switch the 8 and 5 or switch the 4 and 10, then the new statement is still a proportion.

If $8/10 = 4/5$, then $5/10 = 4/8$, or if $8/10 = 4/5$, then $8/4 = 10/5$.

Example 8: If $x/5 = y/4$, find the ratio of x/y.

$$x/5 = y/4$$

Use the switching property of proportions and switch the means positions, the 5 and the y.

$$x/y = 5/4$$

Property 3 (Upside-Down Property): If $a/b = c/d$, then $b/a = d/c$.

Example 9: If $9a = 5b \neq 0$, find the ratio $\frac{a}{b}$.

First, apply the converse of the *Cross Products Property* and obtain the following:

$$\frac{9}{5} = \frac{b}{a} \text{ and } \frac{9}{b} = \frac{5}{a}.$$

Next, proceed in one of the following two ways:

■ Apply *Property 3* to $9/5 = b/a$:

Turn each side upside-down.

$5/9 = a/b$, or $a/b = 5/9$

- Apply *Property 2* to $9/b = 5/a$:

    Switch the 9 and the *a*.

    $a/b = 5/9$

Property 4 (Denominator Addition/Subtraction Property): If $a/b = c/d$, then $(a + b)/b = (c + d)/d$ or $(a - b)/b = (c - d)/d$.

Example 10: If $5/8 = x/y$, then $13/8 = ?$

$$5/8 = x/y$$

Apply *Property 4*. $(5 + 8)/8 = (x + y)/y$

$$13/8 = (x + y)/y$$

Example 11: In Figure 6-1, $AB/BC = 5/8$. Find AC/BC.

Figure 6-1   Using the *Segment Addition Postulate.*

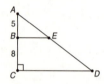

Recall that $AB + BC = AC$ *(Segment Addition Postulate).*

If
$$\frac{AB}{BC} = \frac{5}{8}$$

Then
$$\frac{AB + BC}{BC} = \frac{5 + 8}{8} \quad \text{(Property 4)}$$

So
$$\frac{AC}{BC} = \frac{13}{8}$$

Example 12: A map is scaled so that 3 cm on the map is equal to 5 actual miles. If two cities on the map are 10 cm apart, what is the actual distance the cities are apart?

Let x = the actual distance.

$$\frac{\text{map}}{\text{actual}} = \frac{3}{5} = \frac{10}{x}$$

Apply the *Cross-Products Property.*

$$3x = 50$$
$$x = 16\frac{2}{3}$$

The cities are $16\frac{2}{3}$ miles apart.

Similar Polygons

Two polygons with the same shape are called **similar polygons.** The symbol for "is similar to" is ~. Notice that it is a portion of the "is congruent to" symbol, ≅. When two polygons are similar, these two facts *both* must be true:

■ Corresponding angles are equal.

■ The ratios of pairs of corresponding sides must all be equal.

In Figure 6-2, quadrilateral *ABCD* ~ quadrilateral *EFGH.*

Figure 6-2 Similar quadrilaterals.

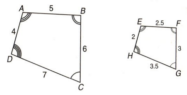

This means: $m \angle A = m \angle E$, $m \angle B = m \angle F$, $m \angle C = m \angle G$, $m \angle D = m \angle H$, and

$$\frac{AB}{EF} = \frac{BC}{FG} = \frac{CD}{GH} = \frac{AD}{EH}$$

It is possible for a polygon to have one of the above facts true without having the other fact true. The following two examples show how that is possible.

In Figure 6-3, quadrilateral QRST is not similar to quadrilateral WXYZ.

Figure 6-3 Quadrilaterals that are not similar to one another.

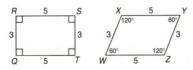

Even though the ratios of corresponding sides are equal, corresponding angles are not equal (90° ≠ 120°, 90° ≠ 60°).

In Figure 6-4, quadrilateral *FGHI* is not similar to quadrilateral *JKLM*.

Figure 6-4   Quadrilaterals that are not similar to one another.

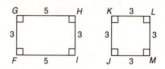

Even though corresponding angles are equal, the ratios of each pair of corresponding sides are not equal $(3/3 \neq 5/3)$.

Example 13: In Figure 6-5, quadrilateral *ABCD* ~ quadrilateral *EFGH*. (a) Find $m \angle E$. (b) Find x.

Figure 6-5   Similar quadrilaterals.

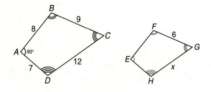

(a) $m \angle E = 90°$ ($\angle E$ and $\angle A$ are corresponding angles of similar polygons, and corresponding angles of similar polygons are equal.)

(b) $9/6 = 12/x$ (If two polygons are similar, then the ratios of each pair of corresponding sides are equal.)

$$9x = 72 \ \textit{(Cross-Products Property)}$$

$$x = 8$$

Similar Triangles

In general, to prove that two polygons are similar, you must show that *all* pairs of corresponding angles are equal and that *all* ratios of pairs of corresponding sides are equal. In triangles, though, this is not necessary.

Postulate 17 (AA Similarity Postulate): If two angles of one triangle are equal to two angles of another triangle, then the triangles are similar.

Example 14: Use Figure 6-6 to show that the triangles are similar.

Figure 6-6 Similar triangles.

$$m \angle B = m \angle E$$

In △ABC,

$$m \angle A + m \angle B + m \angle C = 180°$$
$$m \angle A + 100° + 20° = 180°$$
$$m \angle A = 60°$$

But in △DEF,

$$m \angle D = 60°$$
So, $m \angle A = m \angle D$

By *Postulate 17, the AA Similarity Postulate*, △ABC ~ △DEF. Additionally, because the triangles are now similar,

$$m \angle C = m \angle F$$
and $\dfrac{AB}{DE} = \dfrac{BC}{EF} = \dfrac{AC}{DF}$

Example 15: Use Figure 6-7 to show that △QRS ~ △UTS.

Figure 6-7 Similar triangles.

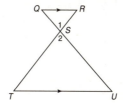

$m \angle 1 = m \angle 2$, because vertical angles are equal.

$m \angle R = m \angle T$ or $m \angle Q = m \angle U$, because if two parallel lines are cut by a transversal, then the alternate interior angles are equal.

So by the *AA Similarity Postulate*, △QRS ~ △UTS.

Example 16: Use Figure 6-8 to show that $\triangle MNO \sim \triangle PQR$.

Figure 6-8 Similar triangles.

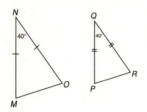

In $\triangle MNO$, $MN = NO$, and in $\triangle PQR$, $PQ = QR$.

> $m \angle M = m \angle O$ and $m \angle P = m \angle R$

(If two sides of a triangle are equal, the angles opposite these sides have equal measures.)

In $\triangle MNO$, $m \angle M + m \angle N + m \angle O = 180°$

In $\triangle PQR$, $m \angle P + m \angle Q + m \angle R = 180°$

Because $m \angle M = m \angle O$ and $m \angle P = m \angle R$,

$$2m \angle M + 40° = 180° \qquad 2m \angle P + 40° = 180°$$
$$2m \angle M = 140° \qquad\qquad 2m \angle P = 140°$$
$$m \angle M = 70° \qquad\qquad m \angle P = 70°$$

So, $m \angle M = m \angle P$, and $m \angle O = m \angle R$. $\triangle MNO \sim \triangle PQR$ (*AA Similarity Postulate*).

Example 17: Use Figure 6-9 to show that $\triangle ABC \sim \triangle DEF$.

Figure 6-9 Similar right triangles.

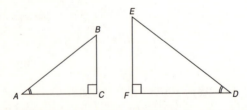

$m \angle C = m \angle F$ (All right angles are equal.)

$m \angle A = m \angle D$ (They are indicated as equal in the figure.)

$\triangle ABC \sim \triangle DEF$ (*AA Similarity Postulate*)

Proportional Parts of Triangles

Consider Figure 6-10 of $\triangle ABC$ with line l parallel to $\overline{AC}$ and intersecting the other two sides at D and E.

Figure 6-10   Deriving the Side-Splitter Theorem.

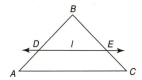

You can eventually prove that $\triangle ABC \sim \triangle DBE$ using the *AA Similarity Postulate*. Because the ratios of corresponding sides of similar polygons are equal, you can show that

$$\frac{AB}{BD} = \frac{BC}{BE}$$

Now use *Property 4*, the *Denominator Subtraction Property.*

$$\frac{AB - DB}{BD} = \frac{BC - BE}{BE}$$

But $AB - DB = AD$, and $BC - BE = CE$ (*Segment Addition Postulate*). With this replacement, you get the following proportion.

$$\frac{AD}{BD} = \frac{CE}{BE}$$

This leads to the following theorem.

Theorem 57 (Side-Splitter Theorem): If a line is parallel to one side of a triangle and intersects the other two sides, it divides those sides proportionally.

Example 18: Use Figure 6-11 to find x.

Figure 6-11 Using the Side-Splitter Theorem.

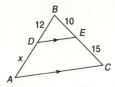

Because $\overline{DE} \parallel \overline{AC}$ in $\triangle ABC$, by *Theorem 57*, you get

$$\frac{x}{12} = \frac{15}{10}$$

$$10x = 180 \ \textit{(Cross-Products Property)}$$

$$x = 18$$

Example 19: Use Figure 6-12 to find x.

Figure 6-12 Using similar triangles.

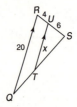

Notice that $\overline{TU}$, x, is *not* one of the segments on either side that $\overline{TU}$ intersects. This means that you *cannot* apply *Theorem 57* to this situation. So what can you do? Recall that with $\overline{TU} \parallel \overline{QR}$, you can show that $\triangle QRS \sim \triangle TUS$. Because the ratios of corresponding sides of similar triangles are equal, you get the following proportion.

$$\frac{QR}{TU} = \frac{RS}{US} = \frac{QS}{TS}$$

$$\frac{20}{x} = \frac{10}{6} = \frac{QS}{TS}$$

$$10x = 120 \ \textit{(Cross-Products Property)}$$

$$x = 12$$

Another theorem involving parts of a triangle is more complicated to prove but is presented here so you can use it to solve problems related to it.

Theorem 58 (Angle Bisector Theorem): If a ray bisects an angle of a triangle, then it divides the opposite side into segments that are proportional to the sides that formed the angle.

In Figure 6-13, $\overline{BD}$ bisects $\angle ABC$ in $\triangle ABC$. By *Theorem 58,* $\dfrac{AD}{DC} = \dfrac{AB}{BC}$.

Figure 6-13   Illustrating the Angle Bisector Theorem.

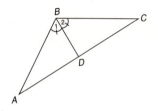

Example 20: Use Figure 6-14 to find x.

Figure 6-14   Using the Angle Bisector Theorem.

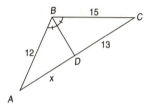

Because $\overline{BD}$ bisects $\angle ABC$ in $\triangle ABC$, you can apply *Theorem 58.*

$$\frac{x}{13} = \frac{12}{15}$$

$$15x = 156 \ \textit{(Cross-Products Property)}$$

$$x = 10\tfrac{2}{5} \text{ or } 10.4$$

Proportional Parts of Similar Triangles

Theorem 59: If two triangles are similar, then the ratio of any two corresponding segments (such as altitudes, medians, or angle bisectors) equals the ratio of any two corresponding sides.

In Figure 6-15, suppose $\triangle QRS \sim \triangle TUV$.

Figure 6-15   Corresponding segments of similar triangles.

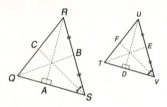

Then, $\dfrac{QR}{TU} = \dfrac{RS}{UV} = \dfrac{QS}{TV}$

Then, according to *Theorem 59,*

$$\frac{\text{length of altitude } \overline{RA}}{\text{length of altitude } \overline{UD}} = \frac{QR}{TU}$$

$$\frac{\text{length of median } \overline{QB}}{\text{length of median } \overline{TE}} = \frac{QR}{TU}$$

$$\frac{\text{length of } \angle \text{bi sector } \overline{CS}}{\text{length of } \angle \text{bi sector } \overline{FV}} = \frac{QR}{TU}$$

Example 21: Use Figure 6-16 and the fact that $\triangle ABC \sim \triangle GHI$ to find x.

Figure 6-16   Proportional parts of similar triangles.

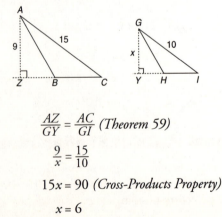

$$\frac{AZ}{GY} = \frac{AC}{GI} \text{ (Theorem 59)}$$

$$\frac{9}{x} = \frac{15}{10}$$

$$15x = 90 \text{ (Cross-Products Property)}$$

$$x = 6$$

Perimeters and Areas of Similar Triangles

When two triangles are similar, the reduced ratio of any two corresponding sides is called the **scale factor** of the similar triangles. In Figure 6-17, $\triangle ABC \sim \triangle DEF$.

Figure 6-17   Similar triangles whose scale factor is 2 : 1.

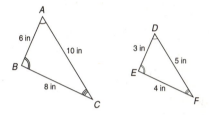

The ratios of corresponding sides are 6/3, 8/4, 10/5. These all reduce to 2/1. It is then said that the scale factor of these two similar triangles is 2 : 1.

The perimeter of $\triangle ABC$ is 24 inches, and the perimeter of $\triangle DEF$ is 12 inches. When you compare the ratios of the perimeters of these similar triangles, you also get 2 : 1. This leads to the following theorem.

Theorem 60: If two similar triangles have a scale factor of $a : b$, then the ratio of their perimeters is $a : b$.

Example 22: In Figure 6-18, $\triangle ABC \sim \triangle DEF$. Find the perimeter of $\triangle DEF$.

Figure 6-18   Perimeter of similar triangles.

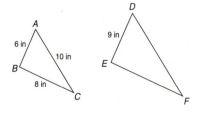

$$\frac{\text{perimeter of } \triangle ABC}{\text{perimeter of } \triangle DEF} = \frac{AB}{DE} \; (\textit{Theorem 60})$$

$$\frac{24}{\text{perimeter of } \triangle DEF} = \frac{6}{9}$$

6(perimeter of $\triangle DEF$) = 216 *(Cross-Products Property)*

perimeter of $\triangle DEF$ = 36 inches

Figure 6-19 shows two similar right triangles whose scale factor is 2 : 3. Because $\overline{GH} \perp \overline{GI}$ and $\overline{JK} \perp \overline{JL}$, they can be considered base and height for each triangle. You can now find the area of each triangle.

Figure 6-19   Finding the areas of similar right triangles whose scale factor is 2 : 3.

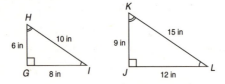

area $\triangle GHI$ = 1/2(6)(8)   area $\triangle JKL$ = 1/2(9)(12)

area $\triangle GHI$ = 24 in^2 area $\triangle JKL$ = 54 in^2

Now you can compare the ratio of the areas of these similar triangles.

$$\frac{\text{area} \triangle GHI}{\text{area} \triangle JKL} = \frac{24}{54}$$

$$= 4/9$$

$$= (2/3)^2$$

This leads to the following theorem:

Theorem 61: If two similar triangles have a scale factor of $a : b$, then the ratio of their areas is $a^2 : b^2$.

Example 23: In Figure 6-20, $\triangle PQR \sim \triangle STU$. Find the area of $\triangle STU$.

Figure 6-20 Using the scale factor to determine the relationship between the areas of similar triangles.

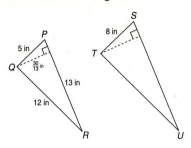

The scale factor of these similar triangles is 5 : 8.

$$\frac{\text{area} \triangle PQR}{\text{area} \triangle STU} = \left(5/8\right)^2 \textit{ (Theorem 61)}$$

$$\text{area} \triangle PQR = \tfrac{1}{2}\left(60/13\right)\left(13\right)$$

$$\text{area} \triangle PQR = 30 \text{ in}^2$$

$$\frac{30}{\text{area} \triangle STU} = \frac{25}{64}$$

$$25\left(\text{area} \triangle STU\right) = 1920 \textit{ (Cross-Products Property)}$$

$$\text{area} \triangle STU = 76\tfrac{4}{5} \text{ in}^2 \text{ or } 76.8 \text{ in}^2$$

Example 24: The perimeters of two similar triangles is in the ratio 3 : 4. The sum of their areas is 75 cm^2. Find the area of each triangle.

If you call the triangles $\triangle_1$ and $\triangle_2$, then

$$\frac{\text{perimeter} \triangle_1}{\text{perimeter} \triangle_2} = \frac{3}{4}$$

According to *Theorem 60*, this also means that the scale factor of these two similar triangles is 3 : 4.

$$\text{Let } 3x = \text{a side in } \triangle_1$$

$$\text{and } 4x = \text{the corresponding side in } \triangle_2.$$

$$\text{Then, } \frac{\text{area} \triangle_1}{\text{area} \triangle_2} = \left(\frac{3x}{4x}\right)^2 \textit{ (Theorem 61)}$$

$$\frac{\text{area} \triangle_1}{\text{area} \triangle_2} = \frac{9x^2}{16x^2}$$

Because the sum of the areas is 75 cm^2, you get

$$\text{area}_{\triangle_1} + \text{area}_{\triangle_2} = 9x^2 + 16x^2$$

$$75 = 25x^2$$

$$3 = x^2$$

$$\text{area}_{\triangle_1} = 9x^2 \qquad\qquad \text{area}_{\triangle_2} = 16x^2$$
$$= 9(3) \qquad\qquad\qquad = 16(3)$$
$$= 27 \text{ cm}^2 \qquad\qquad\quad = 48 \text{ cm}^2$$

Example 25: The areas of two similar triangles are 45 cm^2 and 80 cm^2. The sum of their perimeters is 35 cm. Find the perimeter of each triangle.

Call the two triangles $\triangle_1$ and $\triangle_2$ and let the scale factor of the two similar triangles be $a : b$.

$$\frac{\text{area}_{\triangle_1}}{\text{area}_{\triangle_2}} = \left(\frac{a}{b}\right)^2 \quad \textit{(Theorem 61)}$$

$$\frac{45}{80} = \left(\frac{a}{b}\right)^2$$

Reduce the fraction.

$$\frac{9}{16} = \left(\frac{a}{b}\right)^2$$

Take square roots of both sides.

$$\frac{3}{4} = \frac{a}{b}$$

$a : b$ is the reduced form of the scale factor. 3 : 4 is then the reduced form of the comparison of the perimeters.

$$\text{Let } 3x = \text{perimeter of } \triangle_1$$

$$\text{and } 4x = \text{perimeter of } \triangle_2.$$

$$\text{Then } 3x + 4x = 35 \text{ (The sum of the perimeters is 35 cm.)}$$

$$7x = 35$$

$$x = 5$$

So

$$\text{perimeter}_{\triangle_1} = 3(5) \qquad \text{perimeter}_{\triangle_2} = 4(5)$$
$$= 15 \text{ cm} \qquad\qquad\qquad = 20 \text{ cm}$$

Chapter Checkout

Q&A

1. The ratio of two supplementary angles is 7 to 8. Find the measure of each angle.

2. The following (Figure 6-21) involves △ABC with a line drawn parallel to side AB and intersecting the other two sides. Refer to the figure and compute *x*.

Figure 6-21 Use this figure to compute x.

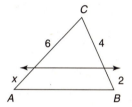

3. Suppose two similar triangles have a scale factor of 3 : 1. If the area of the smaller triangle is 2, then what is the area of the larger triangle?

Answers: 1. 84° and 96° 2. 3 3. 18

Chapter 7

RIGHT TRIANGLES

Chapter Check-In

❑ Deriving the *Pythagorean Theorem* by drawing the altitude to the hypotenuse of a right triangle

❑ Applying the *Pythagorean Theorem* to right triangles

❑ Using the ratios of the sides of an isosceles right triangle to compute an unknown side

❑ Using the ratios of the sides of a 30°-60°-90° right triangle to compute an unknown side

The conclusion of the *Pythagorean Theorem* (see "Pythagorean Theorem and Its Converse" later in this chapter) gives a condition that is both necessary and sufficient for a triangle to be a right triangle. That condition plays a role in mathematics that extends well beyond your current efforts in geometry. It leads to the Pythagorean Identities (not presented here), which have significance in trigonometry, and it is used to define Pythagorean triples, which play a role in number theory. Besides the *Pythagorean Theorem,* this chapter also presents the two triangles "familiar" to almost any trigonometry student: the 45°-45°-90° triangle and the 30°-60°-90° triangle.

Geometric Mean

When a positive value is repeated in either the means or extremes position of a proportion, that value is referred to as a **geometric mean** (or **mean proportional**) between the other two values.

Example 1: Find the geometric mean between 4 and 25.

Let x = the geometric mean.

$$\frac{4}{x} = \frac{x}{25} \text{ (definition of geometric mean)}$$

$$x^2 = 100 \; (\textit{Cross-Products Property})$$

$$x = \sqrt{100}$$

$$x = 10$$

The geometric mean between 4 and 25 is 10.

Example 2: 12 is the geometric mean between 8 and what other value?

Let x = the other value.

$$\frac{8}{12} = \frac{12}{x} \text{ (definition of geometric mean)}$$

$$8x = 144 \; (\textit{Cross-Products Property})$$

$$x = 18$$

The other value is 18.

Altitude to the Hypotenuse

In Figure 7-1, right triangle ABC has altitude $\overline{BD}$ drawn to the hypotenuse $\overline{AC}$.

Figure 7-1   An altitude drawn to the hypotenuse of a right triangle.

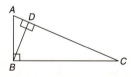

The following theorem can now be easily shown using the *AA Similarity Postulate.*

Theorem 62: The altitude drawn to the hypotenuse of a right triangle creates two similar right triangles, each similar to the original right triangle and similar to each other.

Figure 7-2 shows the three right triangles created in Figure 7-1. They have been drawn in such a way that corresponding parts are easily recognized.

Figure 7-2 Three similar right triangles from Figure 7-1 (not drawn to scale).

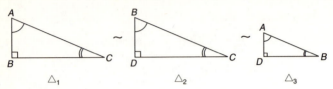

Note that $\overline{AB}$ and $\overline{BC}$ are legs of the original right triangle; $\overline{AC}$ is the hypotenuse in the original right triangle; $\overline{BD}$ is the altitude drawn to the hypotenuse; $\overline{AD}$ is the segment on the hypotenuse touching leg $\overline{AB}$; and $\overline{DC}$ is the segment on the hypotenuse touching leg $\overline{BC}$.

Because the triangles are similar to one another, ratios of all pairs of corresponding sides are equal. This produces three proportions involving geometric means.

(1) $\dfrac{AC}{BC} = \dfrac{BC}{CD} \, (\triangle_1 \sim \triangle_2)$

(2) $\dfrac{AB}{AD} = \dfrac{AC}{AB} \, (\triangle_1 \sim \triangle_3)$

These two proportions can now be stated as a theorem.

Theorem 63: If an altitude is drawn to the hypotenuse of a right triangle, then each leg is the geometric mean between the hypotenuse and its touching segment on the hypotenuse.

(3) $\dfrac{BD}{AD} = \dfrac{CD}{BD} \, (\triangle_2 \sim \triangle_3)$

This proportion can now be stated as a theorem.

Theorem 64: If an altitude is drawn to the hypotenuse of a right triangle, then it is the geometric mean between the segments on the hypotenuse.

Example 3: Use Figure 7-3 to write three proportions involving geometric means.

Figure 7-3 Using geometric means to write three proportions.

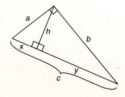

By *Theorem 63,* $\quad \dfrac{c}{a} = \dfrac{a}{x}$ and $\dfrac{c}{b} = \dfrac{b}{y}$

By *Theorem 64,* $\quad \dfrac{x}{b} = \dfrac{b}{y}$

Example 4: Find the values for x and y in Figures 7-4(a) through 7-4(d).

Figure 7-4   Using geometric means to find unknown parts.

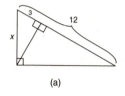

(a)

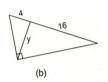

(b)

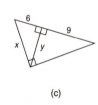

(c)

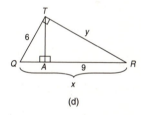

(d)

(a) By *Theorem 63,*

$$12/x = x/3$$
$$x^2 = 36$$
$$x = \sqrt{36}$$
$$x = 6$$

(b) By *Theorem 64,*

$$4/y = y/16$$
$$y^2 = 64$$
$$y = \sqrt{64}$$
$$y = 8$$

(c) By *Theorem 63,*

$$15/x = x/6$$
$$x^2 = 90$$
$$x = \sqrt{90}$$
$$x = (\sqrt{9})(\sqrt{10})$$
$$x = 3\sqrt{10}$$

By *Theorem 64,*

$$6/y = y/9$$
$$y^2 = 54$$
$$y = \sqrt{54}$$
$$y = (\sqrt{9})(\sqrt{6})$$
$$y = 3\sqrt{6}$$

(d) $QA + AR = QR$ *(Segment Addition Postulate)*

$$QA + 9 = x$$

$$QA = x - 9$$

By *Theorem 63*, $x/6 = 6/(x - 9)$

$$x(x - 9) = 36 \text{ (Cross-Products Property)}$$

$$x^2 - 9x = 36$$

$$x^2 - 9x - 36 = 0$$

Factor. $(x - 12)(x + 3) = 0$

$$x - 12 = 0 \text{ or } x + 3 = 0$$

$$x = 12 \text{ or } x = -3$$

Because it represents a length, x cannot be negative, so $x = 12$.

By *Theorem 63*, $x/y = y/9$

Because $x = 12$, from earlier in the problem,

$$12/y = y/9 \text{ (Cross-Products Property)}$$

$$y^2 = 108$$

$$y = \sqrt{108}$$

$$y = \sqrt{(36)\,3}$$

$$y = 6\sqrt{3}$$

Pythagorean Theorem and Its Converse

In Figure 7-5, $\overline{CD}$ is the altitude to hypotenuse $\overline{AB}$.

Figure 7-5 An altitude drawn to the hypotenuse of a right triangle to aid in deriving the *Pythagorean Theorem.*

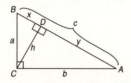

So, by *Theorem 63,*

$c/a = a/x,$ which becomes $a^2 = cx$

and $c/b = b/y,$ which becomes $b^2 = cy$

From the addition property of equations in **algebra,** we get the following equation.

$$a^2 + b^2 = cx + cy$$

By factoring out the c on the right side,

$$a^2 + b^2 = c(x + y)$$

But $x + y = c$ *(Segment Addition Postulate),*

$$a^2 + b^2 = cc \text{ or } a^2 + b^2 = c^2$$

This result is known as the *Pythagorean Theorem.*

Theorem 65 (Pythagorean Theorem): In any right triangle, the sum of the squares of the legs equals the square of the hypotenuse ($\text{leg}^2 + \text{leg}^2 = \text{hypotenuse}^2$). See Figure 7-6 for the parts of a right triangle.

Figure 7-6   Parts of a right triangle.

Example 5: In Figure 7-7, find x, the length of the hypotenuse.

Figure 7-7   Using the *Pythagorean Theorem* to find the hypotenuse of a right triangle.

$$\text{leg}^2 + \text{leg}^2 = \text{hypotenuse}^2$$
$$3^2 + 4^2 = x^2$$
$$9 + 16 = x^2$$
$$25 = x^2$$
$$\sqrt{25} = x$$
$$5 = x$$

Example 6: Use Figure 7-8 to find x.

Figure 7-8   Using the *Pythagorean Theorem* to find the hypotenuse of a right triangle.

$$\text{leg}^2 + \text{leg}^2 = \text{hypotenuse}^2$$
$$5^2 + 7^2 = x^2$$
$$25 + 49 = x^2$$
$$74 = x^2$$
$$\sqrt{74} = x$$

Any three natural numbers, a, b, c, that make the sentence $a^2 + b^2 = c^2$ true are called a Pythagorean triple. Therefore, 3-4-5 is called a Pythagorean triple. Some other values for a, b, and c that will work are 5-12-13 and 8-15-17. Any multiple of one of these triples will also work. For example, using the 3-4-5: 6-8-10, 9-12-15, and 15-20-25 are also Pythagorean triples.

Example 7: Use Figure 7-9 to find x.

Figure 7-9   Using the *Pythagorean Theorem* to find a leg of a right triangle.

If you can recognize that the numbers x, 24, 26 are a multiple of the 5-12-13 Pythagorean triple, the answer for x is quickly found. Because $24 = 2(12)$ and $26 = 2(13)$, then $x = 2(5)$ or $x = 10$. You can also find x by using the *Pythagorean Theorem*.

$$\text{leg}^2 + \text{leg}^2 = \text{hypotenuse}^2$$
$$x^2 + 24^2 = 26^2$$
$$x^2 + 576 = 676$$
$$x^2 = 100$$
$$x = \sqrt{100}$$
$$x = 10$$

Example 8: Use Figure 7-10 to find x.

Figure 7-10   Using the *Pythagorean Theorem* to find the unknown parts of a right triangle.

$$\text{leg}^2 + \text{leg}^2 = \text{hypotenuse}^2$$
$$x^2 + (x + 3)^2 = (x + 6)^2$$
$$x^2 + x^2 + 6x + 9 = x^2 + 12x + 36$$
$$2x^2 + 6x + 9 = x^2 + 12x + 36$$

Subtract $x^2 + 12x + 36$ from both sides.

$$x^2 - 6x - 27 = 0$$

Factor. $$(x - 9)(x + 3) = 0$$
$$x - 9 = 0 \text{ or } x + 3 = 0$$

So, $$x = 9 \text{ or } x = -3$$

But x is a length, so it cannot be negative. Therefore, $x = 9$.

The converse (reverse) of the *Pythagorean Theorem* is also true.

Theorem 66: If a triangle has sides of lengths *a*, *b*, and *c* where *c* is the longest length and $c^2 = a^2 + b^2$, then the triangle is a right triangle with *c* its hypotenuse.

Example 9: Determine if the following sets of lengths could be the sides of a right triangle: (a) 6-5-4, (b) $\sqrt{11}$ - $\sqrt{14}$ -5, (c) 3/4-1-5/4.

(a) Because 6 is the longest length, do the following check.

$$6^2 ? 4^2 + 5^2$$

$$36 ? 16 + 25$$

$$36 \neq 41$$

So 4-5-6 are not the sides of a right triangle.

(b) Because 5 is the longest length, do the following check.

$$5^2 ? \left(\sqrt{11}\right)^2 + \left(\sqrt{14}\right)^2$$

$$25 ? 11 + 14$$

$$25 = 25$$

So √11-√14-5 are sides of a right triangle, and 5 is the length of the hypotenuse.

(c) Because 5/4 is the longest length, do the following check.

$$(5/4)^2 ? (3/4)^2 + (1)^2$$

$$25/16 ? 9/16 + 1$$

$$25/16 = 25/16$$

So 3/4-1-5/4 are sides of a right triangle, and 5/4 is the length of the hypotenuse.

Extension to the Pythagorean Theorem

Variations of *Theorem 66* can be used to classify a triangle as right, obtuse, or acute.

Theorem 67: If *a*, *b*, and *c* represent the lengths of the sides of a triangle, and *c* is the longest length, then the triangle is obtuse if $c^2 > a^2 + b^2$, and the triangle is acute if $c^2 < a^2 + b^2$.

Figures 7-11(a) through 7-11(c) show these different triangle situations and the sentences comparing their sides. In each case, c represents the longest side in the triangle.

right triangle   obtuse triangle   acute triangle

$c^2 = a^2 + b^2$ $c^2 > a^2 + b^2$ $c^2 < a^2 + b^2$

Figure 7-11   The relationship of the square of the longest side to the sum of the squares of the other two sides of a right triangle, an obtuse triangle, and an acute triangle.

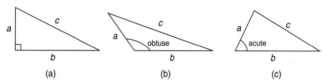

(a)   (b)   (c)

Example 10: Determine whether the following sets of three values could be the lengths of the sides of a triangle. If the values can be the sides of a triangle, then classify the triangle. (a) 16-30-34, (b) 5-5-8, (c) 5-8-15, (d) 4-4-5, (e) 9-12-16, (f) 1-1-$\sqrt{2}$

(Recall the *Triangle Inequality Theorem, Theorem 38,* which states that the longest side in any triangle must be less than the sum of the two shorter sides.)

(a) $$34 ? 16 + 30$$

$$34 < 46 \text{ (So these can be the sides of a triangle.)}$$

$$34^2 ? 16^2 + 30^2$$

$$1156 ? 256 + 900$$

$$1156 = 1156$$

This is a right triangle. Because its sides are of different lengths, it is also a scalene triangle.

(b) $$8 ? 5 + 5$$

$$8 < 10 \text{ (So these can be the sides of a triangle.)}$$

$$8^2 ? 5^2 + 5^2$$

$$64 ? 25 + 25$$

$$64 > 50$$

This is an obtuse triangle. Because two of its sides are of equal measure, it is also an isosceles triangle.

(c) $$15 ? 5 + 8$$

$15 > 13$ (So these *cannot* be the sides of a triangle.)

(d) $$5 ? 4 + 4$$

$5 < 8$ (So these can be the sides of a triangle.)

$$5^2 ? 4^2 + 4^2$$

$$25 ? 16 + 16$$

$$25 < 32$$

This is an acute triangle. Because two of its sides are of equal measure, it is also an isosceles triangle.

(e) $$16 ? 9 + 12$$

$16 < 21$ (So these can be the sides of a triangle.)

$$16^2 ? 9^2 + 12^2$$

$$256 ? 81 + 144$$

$$256 > 225$$

This is an obtuse triangle. Because all sides are of different lengths, it is also a scalene triangle.

(f) $$\sqrt{2} ? 1 + 1$$

$\sqrt{2} < 2$ (So these can be the sides of a triangle.)

$$\left(\sqrt{2}\right)^2 ? 1^2 + 1^2$$

$$2 ? 1 + 1$$

$$2 = 2$$

This is a right triangle. Because two of its sides are of equal measure, it is also an isosceles triangle.

Special Right Triangles

Isosceles right triangle. An **isosceles right triangle** has the characteristic of both the isosceles and the right triangles. It has two equal sides, two equal angles, and one right angle. (The right angle cannot be one of the equal angles or the sum of the angles would exceed 180°.) Therefore, in

Figure 7-12, △ABC is an isosceles right triangle, and the following must always be true.

Figure 7-12 An isosceles right triangle.

$$x + x + 90° = 180° \qquad\qquad △ABC \text{ is isosceles}$$
$$2x = 90° \qquad\qquad AB = BC$$
$$x = 45° \qquad\qquad m\,\angle A = m\,\angle C$$
$$m\,\angle B = 90°$$

The ratio of the sides of an isosceles right triangle is always $1 : 1 : \sqrt{2}$ or $x : x : x\sqrt{2}$ (Figure 7-13).

Figure 7-13 The ratios of the sides of an isosceles right triangle.

Example 11: If one of the equal sides of an isosceles right triangle is 3, what are the measures of the other two sides?

Method 1: Using the ratio $x : x : x\sqrt{2}$ for isosceles right triangles, then $x = 3$, and the other sides must be 3 and $3\sqrt{2}$.

Method 2. Using the *Pythagorean Theorem* and the fact that the legs of this right triangle are equal,

$$\text{leg}^2 + \text{leg}^2 = \text{hypotenuse}^2$$
$$3^2 + 3^2 = \text{hypotenuse}^2$$
$$9 + 9 = \text{hypotenuse}^2$$
$$18 = \text{hypotenuse}^2$$
$$\text{so hypotenuse} = \sqrt{18}$$
$$\text{or hypotenuse} = 3\sqrt{2}$$

The two sides have measures of 3 and $3\sqrt{2}$.

Example 12: If the diagonal of a square is $6\sqrt{2}$, find the length of each of its sides.

Method 1: The diagonal of a square divides it into two congruent isosceles right triangles. Look at Figure 7-14.

Figure 7-14   A diagonal of a square helps create two congruent isosceles right triangles.

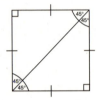

The ratio $x : x : x\sqrt{2}$ for isosceles right triangles can now be applied where $x\sqrt{2} = 6\sqrt{2}$. So $x = 6$, and each side of the square has a measure of 6.

Method 2: Use the *Pythagorean Theorem*. $6\sqrt{2}$ represents the hypotenuse.

$$\text{leg}^2 + \text{leg}^2 = \text{hypotenuse}^2$$
$$(2)\text{leg}^2 = (6\sqrt{2})^2$$
$$(2)\text{leg}^2 = (36)(2)$$
$$(2)\text{leg}^2 = 72$$
$$\text{leg}^2 = 36$$
$$\text{leg} = 6$$

Therefore, each side of the square has a measure of 6.

Example 13: What are the measurements of x, y, and z in Figure 7-15?

Figure 7-15   Finding the unknown parts of this right triangle.

$45° + 90° + x° = 180°$ (The sum of the angles of a triangle = 180°.)

$x = 45°$

Therefore, this is an isosceles right triangle with the ratio of sides $x : x : x\sqrt{2}$. Because one leg is 10, the other must also be 10, and the hypotenuse is $10\sqrt{2}$, so $y = 10$ and $z = 10\sqrt{2}$.

30°-60°-90° right triangle. A **30°-60°-90° right triangle** has a unique ratio of its sides. The ratio of the sides of a 30°-60°-90° right triangle is $1 : \sqrt{3} : 2$ or $x : x\sqrt{3} : 2x$, placed as follows.

> The side opposite 30° is the shortest side and is 1 or x (Figure 7-16).
>
> The side opposite 60° is $\sqrt{3}$ or $x\sqrt{3}$
>
> The side opposite 90° is the longest side (hypotenuse) and is 2 or $2x$.

Figure 7-16   The ratios of the sides of a 30°-60°-90° triangle.

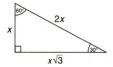

Example 14: If the shortest side of a 30°-60°-90° right triangle is 4, what is the measure of the other two sides?

In Figure 7-17, x is opposite the 30°. The other two sides are then $x\sqrt{3}$ (opposite the 60°) and $2x$ (opposite the 90°). Because the shortest side is 4, $x = 4$. Consequently, the other two sides must be $4\sqrt{3}$ and 2(4), or 8.

Figure 7-17   Using the shortest side of a 30°-60°-90° triangle to find the remaining sides.

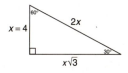

Example 15: If the longer leg of a 30°-60°-90° right triangle is $8\sqrt{3}$, find the length of the hypotenuse.

In Figure 7-18, the shorter leg, x, is opposite the 30°. $x\sqrt{3}$ is the longer leg, and it is opposite the 60°. The hypotenuse is $2x$. Because $x\sqrt{3} = 8\sqrt{3}$, $x = 8$. Because $x = 8$, then $2x = 16$. The hypotenuse is 16.

Figure 7-18   Using the longer leg of a 30°-60°-90° triangle to find the hypotenuse.

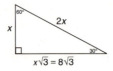

Example 16: Find the length of an altitude in an equilateral triangle with a perimeter of 60 inches.

Figure 7-19 is an equilateral triangle. Each angle has a measure of 60°. If an altitude is drawn, it creates two 30°-60°-90° right triangles. Because the perimeter is 60 inches, and the three sides are equal in measure, then each side is 20 inches $(60 \div 3 = 20)$. The ratio of sides in a 30°-60°-90° right triangle is $x : x\sqrt{3} : 2x$. In this problem, the length 20 inches represents the longest side in the 30°-60°-90° right triangle, so $2x = 20$, or $x = 10$. Because the altitude is the longer leg of the 30°-60°-90° right triangle and its measure is $x\sqrt{3}$, the altitude is $10\sqrt{3}$ inches long.

Figure 7-19   Using the perimeter of an equilateral triangle to find an altitude.

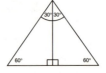

Chapter Checkout

Q&A

1. Determine whether the following sets of three numbers can be the lengths of sides of a triangle. If they can be, then classify the triangle. (a) 1-2-3 (b) 5-12-13 (c) 3-3-5

2. If each side of an equilateral triangle is 2 in long, what is the triangle's altitude?

3. If the hypotenuse of a right triangle is exactly twice as long as the shorter leg, what is the degree measure of the smallest angle in the triangle?

4. Compute the base of an isosceles right triangle with legs 3 in long.

Answers: 1. (a) No triangle (b) right triangle, scalene (c) obtuse triangle, isosceles 2. $\sqrt{3}$ in 3. 30° 4. $3\sqrt{2}$ in.

Chapter 8

CIRCLES

Chapter Check-In

❑ Identifying a circle's special lines and segments

❑ Finding the measure of an inscribed angle

❑ Applying key facts about two chords intersecting inside a circle

❑ Applying key facts about two secants intersecting outside a circle

❑ Computing the arc length and the area of the sector of a circle

As you may already know, the desire to find the circumference and the area of a circle played the key role in the discovery of the irrational number π (approximately 3.14159), but the importance of circles extends well beyond this discovery. Here you will learn about some important angles that are determined using circles, and you will explore some connections between properties of circles and properties of regular polygons. Before these efforts are undertaken, you want to make sure that you are familiar with some conventions involving the terminology and notation.

Parts of Circles

A circle is a special figure, and as such has parts with special names. There are also special angles, lines, and line segments that are exclusive to circles. In this chapter, we shall examine all of them.

- **Circle:** To review, a **circle** is a planar figure consisting of all the points equidistant from a fixed point.

- **Center:** That fixed one point is called the center of the circle. Circles are named by naming the center.

- **Radius:** Any segment with one endpoint at the center of the circle and the other endpoint on the circle is a radius. (The plural of radius is radii.)

- **Chord:** Any segment whose endpoints lie on the circle is a chord.

- **Diameter:** Any chord that passes through the center of the circle is a diameter.

In Figure 8-1:

> circle M
> center M
> radius $\overline{MT}$
> chords $\overline{QR}$ and $\overline{XY}$
> diameter $\overline{XY}$

Figure 8-1   Special points and line segments related to a circle.

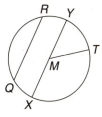

From the definition of radius and diameter, it is clear that all radii of a circle are equal in length and all diameters of a circle are equal in length.

- **Secant:** Any line that contains a chord is a secant.

- **Tangent:** Any line in the same plane as a circle and intersecting the circle at exactly one point is a tangent

- **Point of tangency:** The point where a tangent line intersects a circle is the point of tangency.

In Figure 8-2:

> circle Q
> $\overleftrightarrow{AB}$ is a secant.
> $\overleftrightarrow{CE}$ is a tangent.
> D is the point of tangency for $\overleftrightarrow{CE}$

Figure 8-2 A secant and a tangent to a circle.

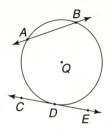

■ **Common tangents:** A line that is tangent to two circles in the same plane is a common tangent.

■ **Internal common tangent:** A common tangent that intersects the segment joining the centers of two circles is an internal common tangent.

■ **External common tangent:** A common tangent that does not intersect the segment joining the centers of two circles is an external common tangent.

In Figure 8-3:

Lines *l* and *m* are common tangents.

l is an internal common tangent.

m is an external common tangent.

Figure 8-3 Internal and external common tangents to circles.

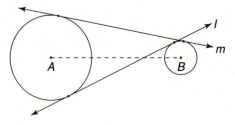

Example 1: Use Figure 8-4 to find each of the following.

(a) radius of circle *O*

(b) chord to circle *P*

(c) diameter of circle *O*

(d) secant to circle P

(e) a tangent to circle O

(f) a common internal tangent to circles O and P

(g) a common external tangent to circles O and P

Figure 8-4   Identifying special lines and line segments related to circles.

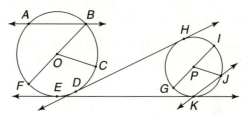

(a) $\overline{OB}$, $\overline{OC}$, or $\overline{OF}$ (b) $\overline{GI}$ or $\overline{JK}$ (c) $\overline{BF}$ (d) $\overleftrightarrow{JK}$

(e) $\overleftrightarrow{KE}$ or $\overrightarrow{DH}$ (f) $\overrightarrow{DH}$ (g) $\overleftrightarrow{EK}$

Central Angles and Arcs

There are several different angles associated with circles. Perhaps the one that most immediately comes to mind is the central angle. It is the central angle's ability to sweep through an arc of 360 degrees that determines the number of degrees usually thought of as being contained by a circle.

Central angles

Central angles are angles formed by any two radii in a circle. The vertex is the center of the circle. In Figure 8-5, $\angle AOB$ is a central angle.

Figure 8-5   A central angle of a circle.

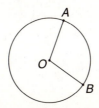

Arcs

An **arc** of a circle is a continuous portion of the circle. It consists of two endpoints and all the points on the circle between these endpoints. The symbol ⌢ is used to denote an arc. This symbol is written over the endpoints that form the arc. There are three types of arcs:

■ **Semicircle:** an arc whose endpoints are the endpoints of a diameter. It is named using three points. The first and third points are the endpoints of the diameter, and the middle point is any point of the arc between the endpoints.

■ **Minor arc:** an arc that is less than a semicircle. A minor arc is named by using only the two endpoints of the arc.

■ **Major arc:** an arc that is more than a semicircle. It is named by three points. The first and third are the endpoints, and the middle point is any point on the arc between the endpoints.

In Figure 8-6, $\overline{AC}$ is a diameter. $\overparen{ABC}$ is a semicircle.

Figure 8-6 A diameter of a circle and a semicircle.

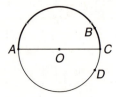

In Figure 8-7, $\overparen{EF}$ is a minor arc of circle P.

Figure 8-7 A minor arc of a circle.

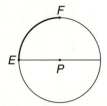

In Figure 8-8, $\overset{\frown}{STU}$ is a major arc of circle Q.

Figure 8-8 A major arc of a circle.

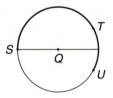

Arcs are measured in three different ways. They are measured in degrees and in unit length as follows:

- **Degree measure of a semicircle:** This is 180°. Its unit length is half of the circumference of the circle.

- **Degree measure of a minor arc:** Defined as the same as the measure of its corresponding central angle. Its unit length is a portion of the circumference. Its length is always less than half of the circumference.

- **Degree measure of a major arc:** This is 360° minus the degree measure of the minor arc that has the same endpoints as the major arc. Its unit length is a portion of the circumference and is always more than half of the circumference.

In this book, $m \overset{\frown}{AB}$ indicates the degree measure of arc AB, $l \overset{\frown}{AB}$ indicates the length of arc AB, and $\overset{\frown}{AB}$ indicates the arc itself.

Example 2: In Figure 8-9, circle O, with diameter $\overline{AB}$, has OB = 6 inches. Find (a) $m \overset{\frown}{AXB}$ and (b) $l \overset{\frown}{AXB}$.

Figure 8-9 Degree measure and arc length of a semicircle.

(a) $\overset{\frown}{AXB}$ is a semicircle. $m\,\overset{\frown}{AXB} = 180°$.

(b) Since $\overset{\frown}{AXB}$ is a semicircle, its length is half of the circumference.

$$l\,\overset{\frown}{AXB} = \tfrac{1}{2}(2\pi r)$$
$$l\,\overset{\frown}{AXB} = \tfrac{1}{2}(2\pi 6 \text{ inches})$$
$$l\,\overset{\frown}{AXB} = 6\pi \text{ inches}$$

Postulate 18 (Arc Addition Postulate): If B is a point on $\overset{\frown}{ABC}$, then $m\,\overset{\frown}{AB} + m\,\overset{\frown}{BC} = m\,\overset{\frown}{ABC}$.

Example 3: Use Figure 8-10 to find $m\,\overset{\frown}{ABC}\left(m\,\overset{\frown}{AB} = 60°, m\,\overset{\frown}{BC} = 150°\right)$.

By *Postulate 18,*
$$m\,\overset{\frown}{ABC} = m\,\overset{\frown}{AB} + m\,\overset{\frown}{BC}$$
$$m\,\overset{\frown}{ABC} = 60° + 150°$$
$$m\,\overset{\frown}{ABC} = 210°$$

Figure 8-10   Using the *Arc Addition Postulate.*

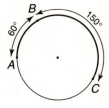

Example 4: Use Figure 8-11 of circle P with diameter $\overline{QS}$ to answer the following.

(a) Find m $\overset{\frown}{RS}$.

(b) Find m $\overset{\frown}{QRS}$.

(c) Find m $\overset{\frown}{QR}$.

(d) Find m $\overset{\frown}{RQS}$.

Figure 8-11   Finding degree measures of arcs.

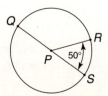

(a) $m\,\overset{\frown}{RS}= 50°$ (The degree measure of a minor arc equals the measure of its corresponding central angle.)

(b) $m\overset{\frown}{QRS} = 180°$ ($\overset{\frown}{QRS}$ is a semicircle.)

(c) $m\,\overset{\frown}{QR} = 130°$

$$m\,\overset{\frown}{QR} + m\,\overset{\frown}{RS} = m\,\overset{\frown}{QRS} \text{ (by } \textit{Postulate 18}\text{)}$$
$$m\,\overset{\frown}{QR} = m\,\overset{\frown}{QRS} - m\,\overset{\frown}{RS}$$
$$m\,\overset{\frown}{QR} = 180° - 50° \text{ (or } 130°\text{)}$$

(d) $m\,\overset{\frown}{RQS} = 310°$ ($\overset{\frown}{RQS}$ is a major arc.) The degree measure of a major arc is 360° minus the degree measure of the minor arc that has the same endpoints as the major arc.

$$m\overset{\frown}{RQS} = 360° - m\,\overset{\frown}{RS}$$
$$m\overset{\frown}{RQS} = 360° - 50° \text{ (or } 310°\text{)}$$

The following theorems about arcs and central angles are easily proven.

Theorem 68: In a circle, if two central angles have equal measures, then their corresponding minor arcs have equal measures.

Theorem 69: In a circle, if two minor arcs have equal measures, then their corresponding central angles have equal measures.

Example 5: Figure 8-12 shows circle O with diameters $\overline{AC}$ and $\overline{BD}$. If $m\angle 1 = 40°$, find each of the following.

(a) $m\,\overset{\frown}{AB}$ (d) $m\angle DOA$

(b) $m\,\overset{\frown}{CD}$ (e) $m\angle 3$

(c) $m\,\overset{\frown}{AD}$ (f) $m\angle 4$

Figure 8-12   A circle with two diameters and a (nondiameter) chord.

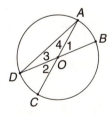

(a) $m\,\overset{\frown}{AB} = 40°$ (The measure of a minor arc equals the measure of its corresponding central angle.)

(b) $m\overparen{CD} = 40°$ (Since vertical angles have equal measures, $m\angle 1 = m\angle 2$. Then the measure of a minor arc equals the measure of its corresponding central angle.)

(c) $m\overparen{AD} = 140°$ (By *Postulate 18*, $m\overparen{AD} + m\overparen{AB} = m\overparen{DAB}$; $\overparen{DAB}$ is a semicircle, so $m\overparen{AD} + 40° = 180°$, or $m\overparen{AD} = 140°$.)

(d) $m\angle DOA = 140°$ (The measure of a central angle equals the measure of its corresponding minor arc.)

(e) $m\angle 3 = 20°$ (Since radii of a circle are equal, $OD = OA$. Since, if two sides of a triangle are equal, then the angles opposite these sides are equal, $m\angle 3 = m\angle 4$. Since the sum of the angles of any triangle equals 180°, $m\angle 3 + m\angle 4 + m\angle DOA = 180°$. By replacing $m\angle 4$ with $m\angle 3$ and $m\angle DOA$ with 140°,

$$m\angle 3 + m\angle 3 + 140° = 180°$$

or $$2\,(m\angle 3) = 40°$$

or $$m\angle 3 = 20°$$

(f) $m\angle 4 = 20°$ (As discussed above, $m\angle 3 = m\angle 4$.)

Arcs and Inscribed Angles

Central angles, as noted previously, are probably the angles most often associated with a circle, but by no means are they the only ones. Angles may be inscribed in the circumference of the circle or formed by intersecting chords and other lines.

■ **Inscribed angle:** In a circle, this is an angle formed by two chords with the vertex on the circle.

■ **Intercepted arc:** Corresponding to an angle, this is the portion of the circle that lies in the interior of the angle together with the endpoints of the arc.

In Figure 8-13, $\angle ABC$ is an inscribed angle and $\overparen{AC}$ is its intercepted arc.

Figure 8-13   An inscribed angle and its intercepted arc.

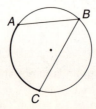

Figure 8-14 shows examples of angles that are *not* inscribed angles.

∠*QRS* is *not* an   ∠*TWV* is *not* an
inscribed angle,   inscribed angle,
since its vertex   since its vertex
is not on the circle.   is not on the circle.

Figure 8-14   Angles that are not inscribed angles.

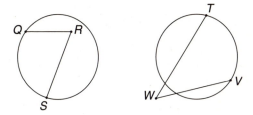

Refer to Figure 8-12 and the example that accompanies it. Notice that *m* ∠3 is exactly half of *m* $\overset{\frown}{AB}$, and *m* ∠4 is half of *m* $\overset{\frown}{CD}$. ∠3 and ∠4 are inscribed angles, and $\overset{\frown}{AB}$ and $\overset{\frown}{CD}$ are their intercepted arcs, which leads to the following theorem.

Theorem 70: The measure of an inscribed angle in a circle equals half the measure of its intercepted arc.

The following two theorems directly follow from *Theorem 70.*

Theorem 71: If two inscribed angles of a circle intercept the same arc or arcs of equal measure, then the inscribed angles have equal measure.

Theorem 72: If an inscribed angle intercepts a semicircle, then its measure is 90°.

Example 6: Find *m* ∠*C* in Figure 8-15.

$$m \angle C = 1/2(m\ \overset{\frown}{BD})\ \textit{(Theorem 70)}$$

$$m \angle C = 1/2(60°)$$

$$m \angle C = 30°$$

Figure 8-15   Finding the measure of an inscribed angle.

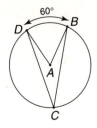

Example 7: Find $m \angle A$ and $m \angle B$ in Figure 8-16.

$$m \angle A = \frac{1}{2}(m\overarc{CD}) \text{ (Theorem 70)}$$
$$m \angle A = \frac{1}{2}(110°)$$
$$m \angle A = 55°$$
$$m \angle B = 55° \text{ (Theorem 71)}$$

Figure 8-16   Two inscribed angles with the same measure.

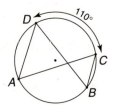

Example 8: In Figure 8-17, $\overline{QS}$ is a diameter. Find $m \angle R$.

$m \angle R = 90°$ *(Theorem 72)*

Figure 8-17   An inscribed angle which intercepts a semicircle.

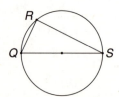

Example 9: In Figure 8-18 of circle O, $m\,\overarc{CD} = 60°$ and $m\,\angle 1 = 25°$.

Find each of the following.

(a) $m\,\angle CAD$

(b) $m\,\overarc{BC}$

(c) $m\,\angle BOC$

(d) $m\,\overarc{AB}$

(e) $m\,\angle ACB$

(f) $m\,\angle ABC$

Figure 8-18   A circle with inscribed angles, central angles, and associated arcs.

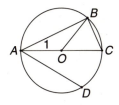

(a) $m\,\angle CAD = \frac{1}{2}(m\,\overarc{CD})$ *(Theorem 70)*

$m\,\angle CAD = \frac{1}{2}(60°)$

$m\,\angle CAD = 30°$

(b) $m\,\angle 1 = \frac{1}{2}(m\,\overarc{BC})$ *(Theorem 70)*

Multiply each side by 2.

$2(m\,\angle 1) = m\,\overarc{BC}$

$2(25°) = m\,\overarc{BC}$

$m\,\overarc{BC} = 50°$

(c) $m\,\angle BOC = 50°$ (The measure of a central angle equals the measure of its corresponding minor arc.)

(d) $m\,\overarc{AB} + m\,\overarc{BC} = m\,\overarc{ABC}$ *(Arc Addition Postulate)*

$m\,\overarc{AB} + 50° = 180°$

$m\,\overarc{AB} = 130°$

(e) $m \angle ACB = \frac{1}{2}(m \widehat{AB})$ *(Theorem 70)*

   $m \angle ACB = \frac{1}{2}(130°)$

   $m \angle ACB = 65°$

(f) $m \angle ABC = 90°$ *(Theorem 72)*

Other Angles Formed by Chords, Secants, and Tangents

Theorem 73: If a tangent and a diameter meet at the point of tangency, then they are perpendicular to one another.

In Figure 8-19, diameter $\overline{AB}$ meets tangent $\overleftrightarrow{CD}$ at B. According to *Theorem 73*, $\overline{AB} \perp \overleftrightarrow{CD}$, which means that $m \angle ABC = 90°$ and $m \angle ABD = 90°$.

Figure 8-19   A tangent to the circle and a diameter of the circle meeting at the point of tangency.

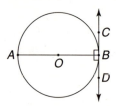

Theorem 74: If a chord is perpendicular to a tangent at the point of tangency, then it is a diameter.

Example 10: *Theorem 74* could be used to find the center of a circle if two tangents to the circle were known. In Figure 8-20, $\overleftrightarrow{MN}$ is tangent to the circle at P, and $\overleftrightarrow{QR}$ is tangent to the circle at S. Use these facts to find the center of the circle.

Figure 8-20   Finding the center of a circle when two tangents to the circle are known.

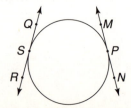

According to *Theorem 74*, if a chord is drawn perpendicular to $\overleftrightarrow{MN}$ at P, it is a diameter, which means that it passes through the center of the circle.

Similarly, if a chord is drawn perpendicular to $\overleftrightarrow{QR}$ at S, it too would be a diameter and pass through the center of the circle. The point where these two chords intersect would then be the center of the circle. See Figure 8-21.

Figure 8-21   Chords drawn perpendicular to tangents to help in finding the center of the circle.

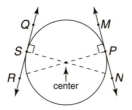

Theorem 75: The measure of an angle formed by two chords intersecting inside a circle is equal to half the sum of the measures of the intercepted arcs associated with the angle and its vertical angle counterpart.

In Figure 8-22, chords $\overline{AC}$ and $\overline{BD}$ intersect inside the circle at E.

By *Theorem 75,*

$$m \angle 1 = \tfrac{1}{2}(m\,\overset{\frown}{AB} + m\,\overset{\frown}{CD})$$

and

$$m \angle 2 = \tfrac{1}{2}(m\,\overset{\frown}{AD} + m\,\overset{\frown}{BC})$$

Figure 8-22   Angles formed by two chords intersecting inside a circle.

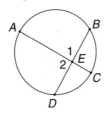

Theorem 76: The measure of an angle formed by a tangent and a chord meeting at the point of tangency is half the measure of the intercepted arc.

In Figure 8-23, chord $\overline{QR}$ and tangent $\overleftrightarrow{TS}$ meet at R. By *Theorem 76,* $m \angle 1 = 1/2\ (m\ \overset{\frown}{QR})$ and $m \angle 2 = \frac{1}{2}\ (m\ \overset{\frown}{QMR})$.

Figure 8-23   A tangent to the circle and a chord meeting at the point of tangency.

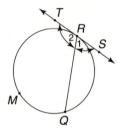

Theorem 77: The measure of an angle formed by two secants intersecting outside a circle is equal to one-half the difference of the measures of the intercepted arcs.

In Figure 8-24, secants $\overleftrightarrow{EF}$ and $\overleftrightarrow{IH}$ intersect at G. According to *Theorem 77, $m \angle 1 - 1/2(m\ \overset{\frown}{EI} - m\ \overset{\frown}{FH})$.*

Figure 8-24   Two secants to the circle meeting outside the circle.

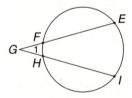

Example 11: Find $m \angle 1$ in Figures 8-25(a) through 8-25(d).

Figure 8-25 Angles formed by intersecting chords, secants, and/or tangents.

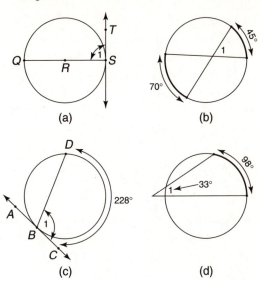

(a)

(b)

(c)

(d)

(a) *m ∠1 = 90° (Theorem 73)*

(b) *m ∠1 = ½(45° + 70°) (Theorem 75)*
     ∠1 = ½(115°)
     m ∠1 = 57 ½°, or 57.5°

(c) *m ∠1 = ½(228°) (Theorem 76)*
     m ∠1 = 114°

(d) *m ∠1 = ½(98° − 33°) (Theorem 77)*
     m ∠1 = ½(65°)
     m ∠1 = 32 ½°, or 32.5°

Example 12: Find the value of *y* in Figures 8-26 (a) through 8-26 (d).

Figure 8-26   Angles formed by intersecting chords, secants, and/or tangents.

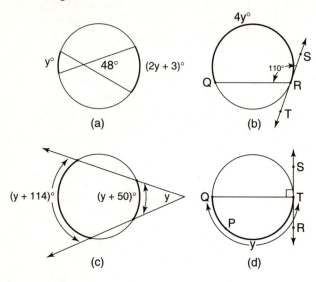

(a) $48 = ½(y + 2y + 3)$ *(Theorem 75)*
        Multiply each side by 2 and simplify.
        $96 = 3y + 3$
        $93 = 3y$
        $31° = y$

(b) $110 = ½(4y)$ *(Theorem 76)*
        $110 = 2y$
        $55° = y$

(c) $y = ½[(y + 114) - (y + 50)]$ *(Theorem 77)*
        Multiply each side by 2 and simplify.
        $2y = y + 114 - y - 50$
        $2y = 64$
        $y = 32°$

(d) $y = 180°$ (According to *Theorem 74*, $\overline{QT}$ is a diameter, which would make $\overset{\frown}{QPT}$ a semicircle.)

Arcs and Chords

In Figure 8-27, circle O has radii $\overline{OA}, \overline{OB}, \overline{OC}$ and $\overline{OD}$. If chords $\overline{AB}$ and $\overline{CD}$ are of equal length, it can be shown that $\triangle AOB \cong \triangle DOC$. This would make $m \angle 1 = m \angle 2$, which in turn would make $m\,\widehat{AB} = m\,\widehat{CD}$. This is stated as a theorem.

Theorem 78: In a circle, if two chords are equal in measure, then their corresponding minor arcs are equal in measure.

The converse of this theorem is also true.

Theorem 79: In a circle, if two minor arcs are equal in measure, then their corresponding chords are equal in measure.

Figure 8-27   A circle with four radii and two chords drawn.

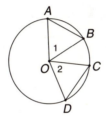

Example 13: Use Figure 8-28 to determine the following. (a) If $AB = CD$, and $m\,\widehat{AB} = 60°$, find $m\,\widehat{CD}$. (b) If $m\,\widehat{EF} = m\,\widehat{GH}$ and $EF = 8$, find GH.

Figure 8-28   The relationship between equality of the measures of (nondiameter) chords and equality of the measures of their corresponding minor arcs.

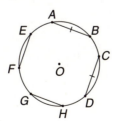

(a) $m\,\widehat{CD} = 60°$ *(Theorem 78)*

(b) $GH = 8$ *(Theorem 79)*

Some additional theorems about chords in a circle are presented below without explanation. These theorems can be used to solve many types of problems.

Theorem 80: If a diameter is perpendicular to a chord, then it bisects the chord and its arcs.

In Figure 8-29, diameter $\overline{UT}$ is perpendicular to chord $\overline{QS}$. By *Theorem 80, QR = RS, m $\widehat{QT}$ = m $\widehat{ST}$,* and $m \widehat{QU}$ = $m \widehat{SU}$.

Figure 8-29   A diameter that is perpendicular to a chord.

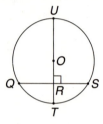

Theorem 81: In a circle, if two chords are equal in measure, then they are equidistant from the center.

In Figure 8-30, if *AB = CD,* then by *Theorem 81, OX = OY.*

Theorem 82: In a circle, if two chords are equidistant from the center of a circle, then the two chords are equal in measure.

In Figure 8-30, if *OX = OY,* then by *Theorem 82, AB = CD.*

Figure 8-30   In a circle, the relationship between two chords being equal in measure and being equidistant from the center.

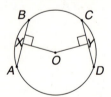

Example 14: Use Figure 8-31 to find *x*.

Figure 8-31 A circle with two minor arcs equal in measure.

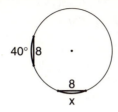

x = 40° *(Theorem 78)*

Example 15: Use Figure 8-32, in which *m* $\widehat{AC}$ = 115°, *m* $\widehat{BD}$ = 115°, and *BD* = *10,* to find *AC.*

Figure 8-32 A circle with two minor arcs equal in measure.

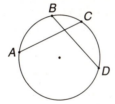

AC = 10 (*Theorem 79*)

Example 16: Use Figure 8-33, in which *AB* = 10, *OA* = 13, and *m* ∠*AOB* = 55°, to find *OM, m* $\widehat{AT}$*,* and *m* $\widehat{SB}$.

Figure 8-33 A circle with a diameter perpendicular to a chord.

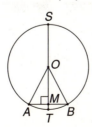

So, $\overline{ST} \perp \overline{AB}$, and $\overline{ST}$ is a diameter. *Theorem 80* says that $AM = BM$. Since $AB = 10$, then $AM = 5$. Now consider right triangle AMO. Since $OA = 13$ and $AM = 5$, OM can be found by using the **Pythagorean Theorem.**

$$OM^2 + AM^2 = AO^2$$

$$OM^2 + 25 = 169$$

$$OM^2 = 144$$

$$OM = \sqrt{144}$$

$$OM = 12$$

Also, *Theorem 80* says that $m\,\overset{\frown}{AT} = m\,\overset{\frown}{TB}$ and $m\,\overset{\frown}{AS} = m\,\overset{\frown}{SB}$. Since $m \angle AOB = 55°$, that would make $m\,\overset{\frown}{AB} = 55°$ and $m\,\overset{\frown}{ASB} = 305°$. Therefore, $m\,\overset{\frown}{AT} = 27\,\tfrac{1}{2}°$ and $m\,\overset{\frown}{SB} = 152\,\tfrac{1}{2}°$.

Example 17: Use Figure 8-34, in which $AB = 8$, $CD = 8$, and $OA = 5$, to find ON.

Figure 8-34   A circle with two chords equal in measure.

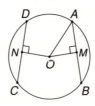

By *Theorem 81*, $ON = OM$. By *Theorem 80*, $AM = MB$, so $AM = 4$. OM can now be found by the use of the *Pythagorean Theorem* or by recognizing a Pythagorean triple. In either case, $OM = 3$. Therefore, $ON = 3$.

Segments of Chords, Secants, and Tangents

In Figure 8-35, chords $\overline{QS}$ and $\overline{RT}$ intersect at P. By drawing $\overline{QT}$ and $\overline{RS}$, it can be proven that $\triangle QPT \sim \triangle RPS$. Because the ratios of corresponding sides of similar triangles are equal, $a/c = d/b$. The *Cross-Products Property* produces $(a)(b) = (c)(d)$. This is stated as a theorem.

Figure 8-35   Two chords intersecting inside a circle.

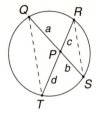

Theorem 83: If two chords intersect inside a circle, then the product of the segments of one chord equals the product of the segments of the other chord.

Example 18: Find x in each of the following figures in Figure 8-36.

Figure 8-36   Two chords intersecting inside a circle.

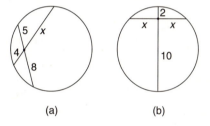

(a)   (b)

(a) By *Theorem 83,* $4(x) = (5)(8)$

$$4x = 40$$
$$x = 10$$

(b) By *Theorem 83,* $(x)(x) = (2)(10)$

$$x^2 = 20$$
$$x = \sqrt{20}$$
$$x = 2\sqrt{5}$$

In Figure 8-37, secant segments $\overline{AB}$ and $\overline{CD}$ intersect outside the circle at E. By drawing $\overline{BC}$ and $\overline{AO}$, it can be proven that $\triangle EBC \sim \triangle EDA$. This makes

$$\frac{EB}{ED} = \frac{EC}{EA}$$

Figure 8-37   Two secant segments intersecting outside a circle.

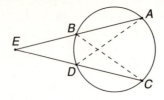

By using the *Cross-Products Property,*

$$(EB)(EA) = (ED)(EC)$$

This is stated as a theorem.

Theorem 84: If two secant segments intersect outside a circle, then the product of the secant segment with its external portion equals the product of the other secant segment with its external portion.

Example 19: Find *x* in each of the following figures in 8-38.

Figure 8-38   More secant segments intersecting outside a circle.

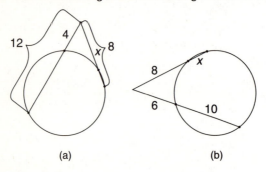

(a)   (b)

(a)   By *Theorem 84,* $8(x) = (12)(4)$

$$8x = 48$$

$$x = 6$$

(b) By *Theorem 84,*
$$(x + 8)(8) = (10 + 6)(6)$$
$$8x + 64 = (16)(6)$$
$$8x + 64 = 96$$
$$8x = 32$$
$$x = 4$$

In Figure 8-39, tangent segment $\overline{AB}$ and secant segment $\overline{BD}$ intersect outside the circle at B. By drawing $\overline{AC}$ and $\overline{AD}$, it can be proven that $\triangle ADB \sim \triangle CAB$. Therefore,

$$\frac{AB}{BC} = \frac{BD}{AB}$$

Figure 8-39 A tangent segment and a secant segment intersecting outside a circle.

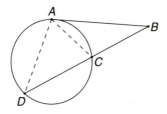

Applying the *Cross-Products Property,*
$$(AB)^2 = (BD)(BC)$$

This is stated as a theorem.

Theorem 85: If a tangent segment and a secant segment intersect outside a circle, then the square of the measure of the tangent segment equals the product of the measures of the secant segment and its external portion.

Also,

Theorem 86: If two tangent segments intersect outside a circle, then the tangent segments have equal measures.

Example 20: Find x in the following figures in 8-40.

Figure 8-40   A tangent segment and a secant segment (or another tangent segment) intersecting outside a circle.

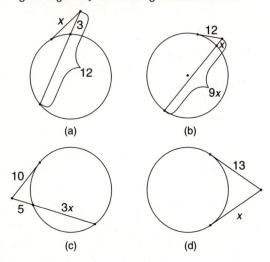

(a)   By *Theorem 85,* $x^2 = (12)(3)$

$$x^2 = 36$$
$$x = \sqrt{36}$$
$$x = 6$$

(b)   By *Theorem 85,* $12^2 = (9x)(x)$
$$144 = 9x^2$$
$$16 = x^2$$
$$\sqrt{16} = x$$
$$4 = x$$

(c)   By *Theorem 85,* $10^2 = (3x + 5)(5)$
$$100 = 15x + 25$$
$$75 = 15x$$
$$5 = x$$

(d)   By *Theorem 86,* $x = 13$

Arc Length and Sectors

Students are often confused by the fact that the arcs of a circle are capable of being measured in more than one way. The best way to avoid that confusion is to remember that arcs possess two properties. They have length as a portion of the circumference, but they also have a measurable curvature, based upon the corresponding central angle.

Arc length

As mentioned earlier in this section, an **arc** can be measured either in degrees or in unit length. In Figure 8-41, $l\,\widehat{AB}$ is a connected portion of the circumference of the circle.

The portion is determined by the size of its corresponding central angle. A proportion will be created that compares a portion of the circle to the whole circle first in degree measure and then in unit length.

Figure 8-41 Determining arc length.

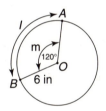

$$\frac{\text{portion of circle}}{\text{whole circle}} = \frac{\text{measure of central angle}}{360°} = \frac{\text{arc length}}{\text{circumference}}$$

$$m°/360° = \tfrac{1}{2}\pi r$$

With the use of this proportion, $\widehat{AB}$ can now be found. In Figure 8-41, the measure of the central angle = 120°, circumference = $2\pi r$, and $r = 6$ inches.

$$\frac{120}{360°} = \frac{l\,\widehat{AB}}{12\pi \text{ inches}}$$

Reduce 120°/360° to ⅓.

$$\frac{1}{3} = \frac{l\,\widehat{AB}}{12\pi \text{ inches}}$$

$$3\left(l\,\widehat{AB}\right) = 12\pi \text{ inches}$$

$$l\,\widehat{AB} = 4\pi \text{ inches}$$

Example 21: In Figure 8-42, $l\,\widehat{AB} = 8\pi$ inches. The radius of the circle is 16 inches. Find $m\angle AOB$.

Figure 8-42   Using the arc length and the radius to find the measure of the associated central angle.

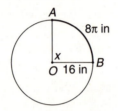

$$x°/360° = \frac{\text{arc length}}{\text{circumference}}$$

$$x°/360° = 8\pi/32\pi$$

Reduce 8π/32π to ¼.

$$x/360° = 1/4$$

$$4x = 360°$$

$$x = 90°$$

So, $m\angle AOB = 90°$

Sector of a circle

A **sector of a circle** is a region bounded by two radii and an arc of the circle.

In Figure 8-43, *OACB* is a sector. $\widehat{ACB}$ is the arc of sector *OACB*. *OADB* is also a sector. $\widehat{ADB}$ is the arc of sector *OADB*. The area of a sector is a portion of the entire area of the circle. This can be expressed as a proportion.

Figure 8-43 A sector of a circle.

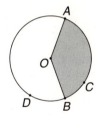

$$\frac{\text{portion of a circle}}{\text{whole circle}} = \frac{\text{measure of central angle}}{360°} = \frac{\text{area of sector}}{\text{area of circle}}$$

$$m°/360° = \text{area of sector}/\pi r^2$$

Example 22: In Figure 8-44, find the area of sector *OACB*.

Figure 8-44 Finding the area of a sector of a circle.

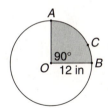

$$\frac{\text{measure of central angle}}{360°} = \frac{\text{area of sector}}{\text{area of circle}}$$

$$\frac{90}{360} = \frac{\text{area of sector}}{\pi(12)^2 \text{in}^2} \quad (\text{area of a circle} = \pi r^2)$$

Reduce $^{90}/_{360}$ to 1/4

$$\frac{1}{4} = \frac{\text{area of sector}}{144\pi \text{ in}^2}$$

$$4(\text{area sector } OACB) = 144\pi \text{ in}^2$$

$$\text{area sector } OACB = 36\pi \text{ in}^2$$

Example 23: In Figure 8-45, find the area of sector *RQTS*.

Figure 8-45　Finding the area of a sector of a circle.

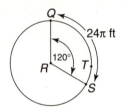

$$\frac{m°}{360°} = \frac{l\,\widehat{QS}}{2\pi r}$$

$$\frac{120}{360} = \frac{24\pi}{2\pi r}$$

Reduce $^{120}/_{360}$ to ⅓ and $\dfrac{24\pi}{2\pi}$ to $^{12}/_{r}$.

$$\frac{1}{3} = \frac{12}{r}$$

$$r = 36$$

The radius of this circle is 36 ft, so the area of the circle is $\pi(36)^2$ or 1296π ft². Therefore,

$$\frac{m°}{360°} = \frac{\text{area of sector}}{\text{area of circle}}$$

$$\frac{120}{360} = \frac{\text{area of sector}}{1296\pi \text{ ft}^2}$$

Reduce $^{120}/_{360}$ to ⅓.

$$\frac{1}{3} = \frac{\text{area of sector}}{1296\pi \text{ ft}^2}$$

$$3(\text{area sector } RQTS) = 1296\pi \text{ ft}^2$$

$$\text{area sector } RQTS = 432\pi \text{ ft}^2$$

Summary of Angle, Segment, Arc Length, and Sector Relationships

Figure 8-46 should help you to visualize at a glance the relationships between the various angles, arcs, and sectors of a circle.

Figure 8-46 Summary of angle, segment, arc length, and sector relationships.

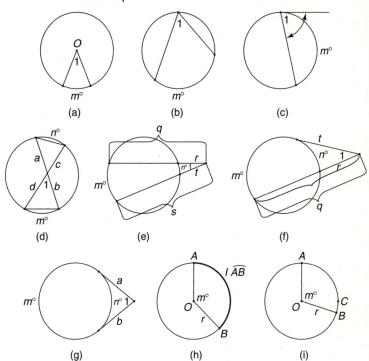

a. $m\angle 1 = m°$

b. $m\angle 1 = \frac{1}{2} m°$

c. $m\angle 1 = \frac{1}{2} m°$

d. $m\angle 1 = \frac{1}{2}(n+m)°$, $ab = cd$

e. $m\angle 1 = \frac{1}{2}(m-n)°$, $qr = st$

f. $m\angle 1 = \frac{1}{2}(m-n)°$, $qr = t^2$

g. $m\angle 1 = 180° - n° = m° - 180°$, $a = b$

h. $l\,\widehat{AB} = m°/360°\,(2\pi r)$, $m\widehat{AB} = m\angle AOB = m°$

i. area of $OACB/\pi r^2 = m°/360°$

Chapter Checkout

Q&A

1. Determine the measure of an inscribed angle whose intercepted arc on the circle is the same as that of a central angle of degree measure 88°.

2. Compute x in the following figure.

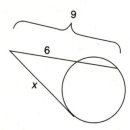

3. Given a circle of radius 3 and a central angle of degree measure 45°, compute the following for the intercepted arc:

(a) the arc length

(b) the area of the sector of the circle bounded by the given central angle and intercepted arc

Answers: 1. 44° 2. $3\sqrt{6}$ 3. (a) $\frac{3\pi}{4}$ (b) $\frac{9\pi}{8}$

Chapter 9
GEOMETRIC SOLIDS

Chapter Check-In

❑ Computing the lateral area and the total surface area for a right prism, a right circular cylinder, a regular pyramid, a right circular cone, and a sphere

❑ Computing the volume of a right prism, a right circular cylinder, a regular pyramid, a right circular cone, and a sphere

Different types of geometric solids affect your life on a continuing basis. For example, tanks of various shapes and sizes are frequently used to store liquids, and from time to time, they require maintenance. To determine how much paint is needed to paint a right circular cylindrical tank, say, or how much oil can be stored in that tank, it is useful to know about the total surface area and volume of a right circular cylinder.

Prisms

Prisms are solids (three-dimensional figures) that, unlike planar figures, occupy space. They come in many shapes and sizes. Every prism has the following characteristics:

■ **Bases:** A prism has two bases, which are congruent polygons lying in parallel planes.

■ **Lateral edges:** The lines formed by connecting the corresponding vertices, which form a sequence of parallel segments.

■ **Lateral faces:** The parallelograms formed by the lateral edges.

A prism is named by the polygon that forms its base, as follows:

■ **Altitude:** A segment perpendicular to the planes of the bases with an endpoint in each plane.

■ **Oblique prism:** A prism whose lateral edges are not perpendicular to the base.

■ **Right prism:** A prism whose lateral edges are perpendicular to the bases. In a right prism, a lateral edge is also an altitude.

Figure 9-1 Different types of prisms.

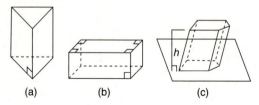

| (a) | (b) | (c) |

In Figure 9-1, prism (a) is a right triangular prism, prism (b) is a right rectangular prism, and prism (c) is an oblique pentagonal prism. The altitude in prism (c) is called *h*.

Right Prisms

In certain prisms, the lateral faces are each perpendicular to the plane of the base (or bases if there is more than one). These are known as a group as right prisms.

Lateral area of a right prism

The lateral area of a right prism is the sum of the areas of all the lateral faces.

Theorem 87: The lateral area, *LA*, of a right prism of altitude *h* and perimeter *p* is given by the following equation.

$$LA_{\text{right prism}} = (p)(h) \text{ units}^2$$

Example 1: Find the lateral area of the right hexagonal prism, shown in Figure 9-2.

Figure 9-2 A right hexagonal prism.

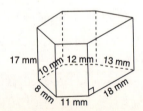

$$LA_{\text{right prism}} = (p)(h) \text{ unit}^2$$
$$= (11 + 18 + 13 + 12 + 10 + 8)(17) \text{ mm}^2$$
$$= (72)(17) \text{ mm}^2$$
$$= 1224 \text{ mm}^2$$

Total area of a right prism

The total area of a right prism is the sum of the lateral area and the areas of the two bases. Because the bases are congruent, their areas are equal.

Theorem 88: The total area, *TA,* of a right prism with lateral area *LA* and a base area *B* is given by the following equation.

$$TA_{\text{right prism}} = LA + 2B \text{ or } TA_{\text{right prism}} = (p)(h) + 2B$$

Example 2: Find the total area of the triangular prism, shown in Figure 9-3.

Figure 9-3   A (right) triangular prism.

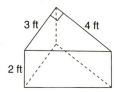

The base of this prism is a right triangle with legs of 3 ft and 4 ft (Figure 9-4).

$$\text{hypotenuse}^2 = 3^2 + 4^2 \text{ (Pythagorean Theorem)}$$
$$\text{hypotenuse}^2 = 9 + 16$$
$$\text{hypotenuse}^2 = 25$$
$$\text{hypotenuse} = \sqrt{25}$$
$$\text{hypotenuse} = 5 \text{ ft}$$

Figure 9-4   The base of the triangular prism from Figure 9-3.

The perimeter of the base is (3 + 4 + 5) ft, or 12 ft.

Because the triangle is a right triangle, its legs can be used as base and height of the triangle.

$$B = \text{Area}_{\text{triangle}} = \frac{1}{2}(b)(h)$$
$$= 1/2(3)(4) \text{ ft}^2$$
$$= 6\text{ft}^2$$

The altitude of the prism is given as 2 ft. Therefore,

$$TA_{\text{right prism}} = LA + 2B \text{ units}^2$$
$$= (p)(h) + 2B \text{ units}^2$$
$$= (12 \text{ ft})(2 \text{ ft}) + (2)(6) \text{ ft}^2$$
$$= 24 \text{ ft}^2 + 12 \text{ ft}^2$$
$$= 36\text{ft}^2$$

Interior space of a solid

Lateral area and total area are measurements of the surface of a solid. The interior space of a solid can also be measured.

A **cube** is a square right prism whose lateral edges are the same length as a side of the base; see Figure 9-5.

Figure 9-5   A cube.

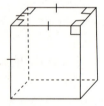

The **volume** of a solid is the number of cubes with unit edge necessary to entirely fill the interior of the solid. In Figure 9-6, the right rectangular prism measures 3 inches by 4 inches by 5 inches.

Figure 9-6   Volume of a right rectangular prism.

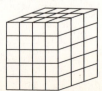

This prism can be filled with cubes 1 inch on each side, which is called a **cubic inch.** The top layer has 12 such cubes. Because the prism has 5 such layers, it takes 60 of these cubes to fill this solid. Thus, the volume of this prism is 60 cubic inches.

Theorem 89: The volume, V, of a right prism with a base area B and an altitude h is given by the following equation.

$$V_{\text{right prism}} = (B)(h) \text{ unit}^3$$

Example 3: Figure 9-7 is an isosceles trapezoidal right prism. Find (a) LA (b) TA and (c) V.

Figure 9-7 An isosceles trapezoidal right prism.

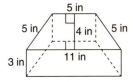

(a) $LA_{\text{right prism}} = (p)(h) \text{ units}^2$

(***Note:*** The h refers to the altitude of the prism, not the height of the trapezoid.)

$$= (5 + 5 + 5 + 11)(3) \text{ in}^2$$
$$= (26)(3) \text{ in}^2$$
$$= 78 \text{ in}^2$$

(b) $B = \text{Area}_{\text{trapezoid}}$ $TA_{\text{right prism}} = LA + 2B \text{ units}^2$

   $B = 1/2(5 + 11)(4) \text{ in}^2$ $= 78 + 2(32)$

   $B = 1/2(16)(4) \text{ in}^2$ $= 78 + 64$

   $B = 32 \text{in}^2$ $= 142 \text{ in}^2$

(c) $V_{\text{right prism}} = (B)(h) \text{ units}^3$

(***Note:*** The h refers to the altitude of the prism, not the height of the trapezoid.)

$$= (32)(3) \text{ in}^3$$
$$= 96 \text{ in}^3$$

Right Circular Cylinders

A prism-shaped solid whose bases are circles is a **cylinder.** If the segment joining the centers of the circles of a cylinder is perpendicular to the planes of the bases, the cylinder is a **right circular cylinder.** In Figure 9-8, cylinder (a) is a right circular cylinder and cylinder (b) is an oblique circular cylinder.

Figure 9-8 Different types of circular cylinders.

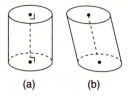

(a)   (b)

Lateral area, total area, and volume for right circular cylinders are found in the same way as they are for right prisms.

If a cylinder is pictured as a soup can, its lateral area is the area of the label. If the label is carefully peeled off, the label becomes a rectangle, as shown in Figure 9-9.

Figure 9-9 The lateral area of a cylinder.

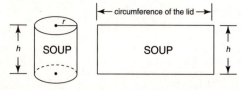

The area of the label is the area of a rectangle with a height the same as the altitude of the can and a base the same as the circumference of the lid of the can.

Theorem 90: The lateral area, *LA,* of a right circular cylinder with a base circumference *C* and an altitude *h* is given by the following equation.

$$LA_{\text{right circular cylinder}} = (C)(h) \text{ unit}^2$$
$$= (2\pi r)(h) \text{ unit}^2$$

Theorem 91: The total area, *TA*, of a right circular cylinder with lateral area *LA* and a base area *B* is given by the following equation.

$$TA_{\text{right circular cylinder}} = LA + 2B \text{ unit}^2$$
$$= (2\pi r)(h) + 2\pi r^2 \text{ units}^2$$
$$= 2\pi r(h + r) \text{ units}^2$$

Theorem 92: The volume of a right circular cylinder, *V*, with a base area *B* and altitude *h* is given by the following equation.

$$V_{\text{right circular cylinder}} = (B)(h) \text{ units}^3$$
$$= (\pi r^2)(h) \text{ units}^3$$

Example 4: Figure 9-10 is a right circular cylinder; find (a) *LA* (b) *TA* and (c) *V*.

Figure 9-10   Finding the lateral area, total area, and volume of a right circular cylinder.

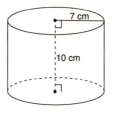

(a) $LA_{\text{right circular cylinder}} = (C)(h) \text{ units}^2$
$$= (2)(\pi)(7)(10) \text{ cm}^2$$
$$= 140\pi \text{ cm}^2$$

(b) $TA_{\text{right circular cylinder}} = LA + 2B \text{ units}^2$
$$= 140\pi + 2(\pi)(7)^2 \text{ cm}^2$$
$$= 140\pi + 98\pi \text{ cm}^2$$
$$= 238\pi \text{ cm}^2$$

(C) $V_{\text{right circular cylinder}} = (B)(h) \text{ units}^3$
$$= (\pi)(7)^2(10) \text{ cm}^3$$
$$= (49\pi)(10) \text{ cm}^3$$
$$= 490\pi \text{ cm}^3$$

Pyramids

A **pyramid** is a solid that has the following characteristics.

■ It has one **base**, which is a polygon.

■ The vertices of the base are each joined to a point, not in the plane of the base. This point is called the **vertex** of the pyramid.

■ The triangular sides, all of which meet at the vertex, are its **lateral faces.**

■ The segments where the lateral faces intersect are **lateral edges.**

■ The perpendicular segment from the vertex to the plane of the base is the **altitude** of the pyramid.

Regular Pyramids

A **regular pyramid** is a pyramid whose base is a regular polygon and whose lateral edges are all equal in length. A pyramid is named by its base. Figure 9-11 shows some examples of regular pyramids.

Figure 9-11 Some different types of regular pyramids.

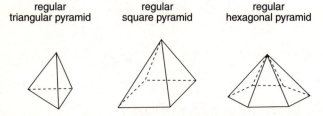

regular
triangular pyramid

regular
square pyramid

regular
hexagonal pyramid

The lateral faces of a regular pyramid are congruent isosceles triangles. The altitude of any of these triangles is the **slant height** of the regular pyramid. Figure 9-12 is a square pyramid.

Figure 9-12 A square pyramid.

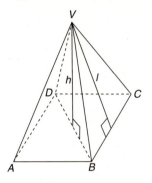

Square *ABCD* is its base.

V is the vertex.

Triangle *VAB* is a lateral face.

VA is a lateral edge.

h is the altitude.

l is the slant height.

Pyramids also have a lateral area, total area, and volume.

Theorem 93: The lateral area, *LA,* of a regular pyramid with slant height *l* and base perimeter *p* is given by the following equation.

$$LA_{\text{regular pyramid}} = 1/2(p)(l) \text{ units}^2$$

Example 5: Find the lateral area of the square pyramid, shown in Figure 9-13.

Figure 9-13 Finding the lateral area, total area, and volume of a square pyramid.

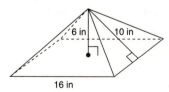

$$P_{square} = 4 \times side$$
$$P_{square} = 4 \times 16 \text{ in}$$
$$P_{square} = 64 \text{ in}$$
$$LA_{regular\ pyramid} = 1/2(p)(l) \text{ units}^2$$
$$= 1/2(64)(10) \text{ in}^2$$
$$= 320 \text{ in}^2$$

Because a pyramid has only one base, its total area is the sum of the lateral area and the area of its base.

Theorem 94: The total area, *TA*, of a regular pyramid with lateral area *LA* and base area *B* is given by the following equation.

$$TA_{regular\ pyramid} = LA + B \text{ units}^2$$
$$= 1/2(p)(l) + B \text{ units}^2$$

Example 6: Find the total area of the regular pyramid shown in Figure 9-13.

The base of the regular pyramid is a **square.** $A_{square} = (side)^2$. Therefore, *B* = 16^2 in^2, or *B* = 256 in^2.

$$TA_{regular\ pyramid} = LA + B \text{ units}^2$$

From the previous example,

$$LA = 320 \text{ in}^2.$$
$$TA = 320 + 256 \text{ in}^2$$
$$= 576 \text{ in}^2$$

Theorem 95: The volume, *V*, of a regular pyramid with base area *B* and altitude *h* is given by the following equation.

$$V_{regular\ pyramid} = 1/3(B)(h) \text{ units}^3$$

Example 7: Find the volume of the regular pyramid shown in Figure 9-13.

From the previous example, *B* = 256 in^2. The figure indicates that *h* = 6 in.

$$V_{regular\ pyramid} = 1/3(B)(h) \text{ units}^3$$
$$= 1/3(256)(6) \text{ in}^3$$
$$= 512 \text{ in}^3$$

Right Circular Cones

A **right circular cone** is similar to a regular pyramid except that its base is a circle. The vocabulary and equations pertaining to the right circular cone are similar to those for the regular pyramid. Refer to Figure 9-14 for the vocabulary regarding right circular cones.

Figure 9-14   A right circular cone.

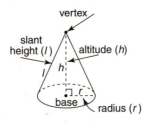

Theorem 96: The lateral area, *LA,* of a right circular cone with base circumference *C* and slant height *l* is given by the following equation.

$$LA_{\text{right circular cone}} = 1/2(C)(l) \text{ units}^2$$
$$= 1/2(2\pi)(r)(l) \text{ units}^2$$
$$= \pi rl \text{ units}^2$$

Theorem 97: The total area, *TA,* of a right circular cone with lateral area *LA* and base area *B* is given by the following equation.

$$TA_{\text{right circular cone}} = LA + B \text{ units}^2$$
$$= \pi rl + \pi r^2 \text{ units}^2$$
$$= \pi r(l + r) \text{ units}^2$$

Theorem 98: The volume, *V,* of a right circular cone with base area *B* and altitude *h* is given by the following equation.

$$V_{\text{right circular cone}} = 1/3(B)(h) \text{ units}^3$$
$$= 1/3(\pi r^2)(h) \text{ units}^3$$

Example 8: Figure 9-15 is a right circular cone; find (a) *LA* (b) *TA* and (c) *V.*

Figure 9-15 Finding the lateral area, total area, and volume of a right circular cone.

(a) The slant height, radius, and altitude of a right circular cone form a right triangle, as shown in Figure 9-16.

Figure 9-16 The right triangle formed by the slant height, radius, and altitude of a right circular cone.

$$l^2 = 8^2 + 6^2 \text{ (Pythagorean Theorem)}$$
$$l^2 = 64 + 36$$
$$l^2 = 100$$
$$l = \sqrt{100}$$
$$l = 10 \text{ cm}$$
$$LA_{\text{right circular cone}} = \pi r l \text{ units}^2$$
$$= \pi(6)(10) \text{ cm}^2$$
$$= 60\pi \text{ cm}^2$$

(b)
$$TA_{\text{right circular cone}} = LA + B \text{ units}^2$$
$$= 60\pi + \pi(6)^2 \text{ cm}^2$$
$$= 60\pi + 36\pi \text{ cm}^2$$
$$= 96\pi \text{ cm}^2$$

(c)
$$V_{\text{right circular cone}} = 1/3 \, (B)(h) \text{ units}^3$$
$$= 1/3(36\pi)(8) \text{ cm}^3$$
$$= 96\pi \text{ cm}^3$$

Spheres

A **sphere** is the set of all points in space that are equidistant from a fixed point (the *center*). That distance is the **radius of the sphere.** Because a sphere has no bases, its area is referred to as its **surface area.**

Theorem 99: The surface area, S, of a sphere with radius r is given by the following equation:

$$S_{sphere} = 4\pi r^2 \text{ units}^2$$

Theorem 100: The volume of a sphere, V, with radius r is given by the following equation:

$$V_{sphere} = \frac{4}{3}\pi r^3 \text{ units}^3$$

Example 9: Figure 9-17 represents a sphere with radius r. If $r = 9$ cm, find (a) S and (b) V.

Figure 9-17 Finding the surface area and volume of a sphere.

(a) $S_{sphere} = 4\pi r^2 \text{ units}^2$

$= 4\pi(9)^2 \text{ cm}^2$

$= 4\pi(81) \text{ cm}^2$

$= 324\pi \text{ cm}^2$

(b) $V_{sphere} = 4/3\pi r^3 \text{ units}^3$

$= 4/3\pi(9)^3 \text{ cm}^3$

$= 4/3 \ \pi(729) \text{ cm}^3$

$= 972\pi \text{ cm}^3$

See Figure 9-18 for a review of geometric solids.

Figure 9-18   Summary of formulas concerning geometric solids.

Name	Example Figure	Lateral Area	Total Area	Volume
right prism	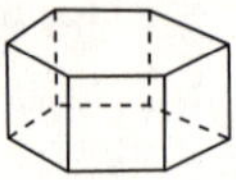 (a)	ph	L.A. $+ 2B$ $= ph + 2B$	Bh
right circular cylinder	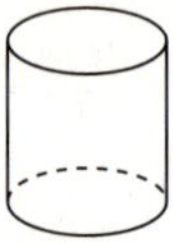 (b)	$Ch = 2\pi rh$	L.A. $+ 2B$ $= 2\pi rh + 2\pi r^2$ $= 2\pi r(h + r)$	Bh
regular pyramid	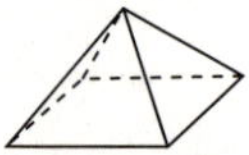 (c)	$1/2pl$	L.A. $+ B$ $= 1/2pl + B$	$1/3Bh$
right circular cone	 (d)	$1/2Cl$ $= 1/2(2\pi r)l$ $= \pi rl$	L.A. $+ B$ $= \pi rl + \pi r^2$ $= \pi r(l + r)$	$1/3Bh$ $= 1/3\pi r^2h$
sphere	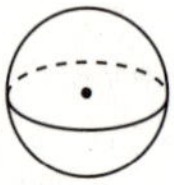 (e)	none	$S = 4\pi r^2$	$4/3\pi r^3$

C = circumference of a base

r = radius of a circle

p = perimeter of a base

B = area of a base

h = altitude

l = slant height

S = surface area of a sphere

Chapter Checkout

Q&A

1. Find (a) the lateral area, (b) the total area, and (c) the volume of a right circular cylinder with radius 4 in. and height 10 in.

2. Find (a) the lateral area, (b) the total area, and (c) the volume of a regular pyramid with altitude 12 in. and a square base with side 6 in. (***Hint:*** Use the *Pythagorean Theorem* to compute the slant height.)

3. Find (a) the surface area and (b) the volume of a sphere of radius $r = 2$ in.

Answers: 1. (a) 80π in^2 (b) 112π in^2 (c) 160π in^3 2. (a) $36\sqrt{17}$ in^2 (b) $(36\sqrt{17} + 36)$ in^2 (c) 144 in^3 3. (a) 16π in^2 (b) $\frac{32}{3}\pi$ in^3

Chapter 10

COORDINATE GEOMETRY

Chapter Check-In

❑ Computing the distance between two points in a plane

❑ Finding the midpoint and the slope of a line segment connecting two points

❑ Determining whether lines are parallel or perpendicular or neither

❑ Determining the slope and the intercepts of a line

❑ Determining an equation for a straight line

The coordinate system that you study in this chapter has axes that are perpendicular to one another, so the system is called a **rectangular coordinate system.** It is also referred to as the Cartesian coordinate system, named after Rene Descartes, a seventeenth-century French mathematician/philosopher. This coordinate system provides a powerful tool for visualizing the functions studied in algebra, trigonometry, calculus, and higher mathematics. As usual, the place to start is by defining key terminology.

Points and Coordinates

Every point in space can be assigned three numbers with respect to a starting point. Those three numbers allow us to distinguish any point from any other point in space. Fortunately for you, we are not dealing here with three dimensions, but only with two.

- **Coordinates of a point:** Each point on a number line is assigned a number. In the same way, each point in a plane is assigned a pair of numbers.

- **x-axis and y-axis:** To locate points in a plane, two perpendicular lines are used—a horizontal line called the x-axis and a vertical line called the y-axis.

- **Origin:** The point of intersection of the *x*-axis and *y*-axis.

- **Coordinate plane:** The *x*-axis, *y*-axis, and all the points in the plane they determine.

- **Ordered pairs:** Every point in a coordinate plane is named by a pair of numbers whose order is important; these numbers are written in parentheses and separated by a comma.

- **x-coordinate:** The number to the left of the comma in an ordered pair is the *x*-coordinate of the point and indicates the amount of movement along the *x*-axis from the origin. The movement is to the right if the number is positive and to the left if the number is negative.

- **y-coordinate:** The number to the right of the comma in an ordered pair is the *y*-coordinate of the point and indicates the amount of movement perpendicular to the *x*-axis. The movement is above the *x*-axis if the number is positive and below the *x*-axis if the number is negative.

Note: The coordinates [ordered pair] for the origin is (0, 0).

The *x*-axis and *y*-axis separate the coordinate plane into four regions called **quadrants.** (See Figure 10-1.) The upper right quadrant is quadrant 1; the upper left quadrant is quadrant II; the lower left quadrant is quadrant III; and the lower right quadrant is quadrant IV. Notice the following:

- In quadrant I, *x* is always positive and *y* is always positive.

- In quadrant II, *x* is always negative and *y* is always positive.

- In quadrant III, *x* is always negative and *y* is always negative.

- In quadrant IV, *x* is always positive and *y* is always negative.

Figure 10-1 The coordinate axes separate the plane into four quadrants.

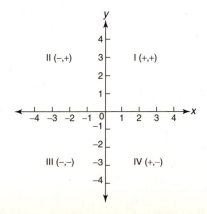

The point associated with an ordered pair of real numbers is called the **graph** of the ordered pair.

Example 1: Identify the points *A, B, C, D, E,* and *F* on the coordinate graph in Figure 10-2.

Figure 10-2 Finding the coordinates of specific points in the plane.

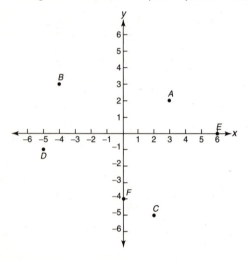

$A(3,2)$, $B(-4,3)$, $C(2,-5)$, $D(-5,-1)$, $E(6,0)$, and $F(0,-4)$

Example 2: Rectangle *ABCD* has coordinates as follows: $A(-5,2)$, $B(8,2)$, and $C(8,-4)$. Find the coordinates of *D*.

A graph is helpful in solving this problem. Refer to Figure 10-3. The coordinates of *D* must be $(-5,-4)$.

Figure 10-3 Finding the coordinates of the fourth vertex of a rectangle.

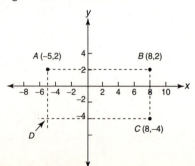

Example 3: Use Figure 10-3 to find the following distances: (a) from A to B (called AB) and (b) from B to C (called BC).

(a) $AB = 8 - (-5)$ and   (b) $BC = 2 - (-4)$ *(Postulate 7)*

     $AB = 13$ $BC = 6$

Distance Formula

In Figure 10-4, A is (2, 2), B is (5, 2), and C is (5, 6).

Figure 10-4   Finding the distance from A to C.

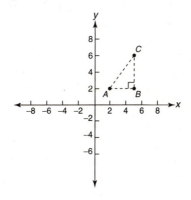

To find AB or BC, only simple subtracting is necessary.

     $AB = 5 - 2$ and $BC = 6 - 2$

     $AB = 3$ $BC = 4$

To find AC, though, simply subtracting is not sufficient. Triangle ABC is a right triangle with $\overline{AC}$ the hypotenuse. Therefore, by the *Pythagorean Theorem*,

$$\overline{AC}^2 = AB^2 + BC^2$$
$$\overline{AC} = \sqrt{AB^2 + BC^2}$$
$$\overline{AC} = \sqrt{3^2 + 4^2}$$
$$\overline{AC} = \sqrt{9 + 16}$$
$$\overline{AC} = \sqrt{25}$$
$$\overline{AC} = 5$$

If A is represented by the ordered pair (x_1, y_1) and C is represented by the ordered pair (x_2, y_2), then $AB = (x_2 - x_1)$ and $BC = (y_2 - y_1)$.

Then

$$AC = \sqrt{(x_2 - x_1)^2 + (y_2 - y_1)^2}$$

This is stated as a theorem.

Theorem 101: If the coordinates of two points are (x_1, y_1) and (x_2, y_2), then the distance, d, between the two points is given by the following formula *(Distance Formula)*.

$$d = \sqrt{(x_2 - x_1)^2 + (y_2 - y_1)^2}$$

Example 4: Use the *Distance Formula* to find the distance between the points with coordinates (–3, 4) and (5, 2).

Let $(-3, 4) = (x_1, y_1)$ and $(5, 2) = (x_2, y_2)$. Then

$$d = \sqrt{(5 - (-3))^2 + (2 - 4)^2}$$
$$d = \sqrt{(8)^2 + (2)^2}$$
$$d = \sqrt{64 + 4}$$
$$d = \sqrt{68}$$
$$d = \sqrt{(4)(17)}$$
$$d = 2\sqrt{17}$$

Example 5: A triangle has vertices $A(12,5)$, $B(5,3)$, and $C(12, 1)$. Show that the triangle is isosceles.

By the *Distance Formula,*

$$AB = \sqrt{(5 - 12)^2 + (3 - 5)^2} \qquad BC = \sqrt{(12 - 5)^2 + (1 - 3)^2}$$
$$AB = \sqrt{(7^2) + (-2)} \qquad BC = \sqrt{7^2 + (-2)^2}$$
$$AB = \sqrt{49 + 4} \qquad BC = \sqrt{49 + 4}$$
$$AB = \sqrt{53} \qquad BC = \sqrt{53}$$

Because $AB = BC$, triangle ABC is isosceles.

Midpoint Formula

Numerically, the midpoint of a segment can be considered to be the average of its endpoints. This concept helps in remembering a formula for finding the midpoint of a segment given the coordinates of its endpoints. Recall that the average of two numbers is found by dividing their sum by two.

Theorem 102: If the coordinates of *A* and *B* are (x_1, y_1) and (x_2, y_2) respectively, then the midpoint, *M*, of *AB* is given by the following formula (*Midpoint Formula*).

$$M = \left(\frac{x_1 + x_2}{2}, \frac{y_1 + y_2}{2} \right)$$

Example 6: In Figure 10-5, *R* is the midpoint between $Q(-9, -1)$ and $T(-3, 7)$. Find its coordinates and use the *Distance Formula* to verify that it is in fact the midpoint of $\overline{QT}$.

Figure 10-5 Finding the coordinates of the midpoint of a line segment.

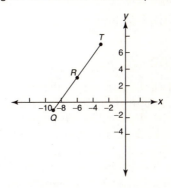

By the *Midpoint Formula*,
$$R = \left(\frac{-9 + -3}{2}, \frac{-1 + 7}{2} \right)$$
$$R = \left(\frac{-12}{2}, \frac{6}{2} \right)$$
$$R = (-6, 3)$$

By the *Distance Formula*,

$$QR = \sqrt{\left(-6 - (-9)\right)^2 + \left(3 - (-1)\right)^2} \qquad TR = \sqrt{\left(-6 - (-3)\right)^2 + (3 - 7)^2}$$
$$QR = \sqrt{3^2 + 4^2} \qquad\qquad\qquad TR = \sqrt{(-3)^2 + (-4)^2}$$
$$QR = \sqrt{9 + 16} \qquad\qquad\qquad TR = \sqrt{9 + 16}$$
$$QR = \sqrt{25} \qquad\qquad\qquad\quad TR = \sqrt{25}$$
$$QR = \sqrt{5} \qquad\qquad\qquad\quad\; TR = \sqrt{25}$$

Because $QR = TR$ and Q, T, and R are collinear, R is the midpoint of $\overline{QT}$.

Example 7: If the midpoint of $\overline{AB}$ is (−3, 8) and *A* is (12, −1), find the coordinates of *B*.

Let the coordinates of *B* be (*x, y*). Then by the *Midpoint Formula,*

$$(-3,8) = \left(\frac{12 + x}{2}, \frac{-1 + y}{2} \right)$$

$$-3 = \frac{12 + x}{2} \quad \text{and} \quad 8 = \frac{-1 + y}{2}$$

Multiply each side of each equation by 2.

$$-6 = 12 + x \quad \text{and} \quad 16 = -1 + y$$
$$-18 = x \quad \text{and} \quad 17 = y$$

Therefore, the coordinates of *B* are (−18, 17).

Slope of a Line

The **slope of a line** is a measurement of the steepness and direction of a nonvertical line. When a line rises from left to right, the slope is a positive number. Figure 10-6(a) shows a line with a positive slope. When a line falls from left to right, the slope is a negative number. Figure 10-6(b) shows a line with a negative slope. The *x*-axis or any line parallel to the *x*-axis has a slope of zero. Figure 10-6(c) shows a line whose slope is zero. The *y*-axis or any line parallel to the *y*-axis has no defined slope. Figure 10-6(d) shows a line with an undefined slope.

Figure 10-6   Different possibilities for slope of a line.

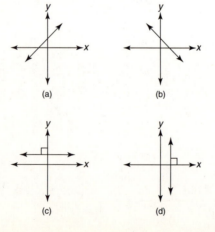

(a)   (b)

(c)   (d)

If m represents the slope of a line and A and B are points with coordinates (x_1, y_1) and (x_2, y_2) respectively, then the slope of the line passing through A and B is given by the following formula.

$$m = \frac{y_2 - y_1}{x_2 - x_1}, \text{ if } x_2 \neq x_1$$

A and B cannot be points on a vertical line, so x_1 and x_2 cannot be equal to one another. If $x_1 = x_2$, then the line is vertical and the slope is undefined.

Example 8: Use Figure 10-7 to find the slopes of lines a, b, c, and d.

Figure 10-7: Finding the slopes of specific lines.

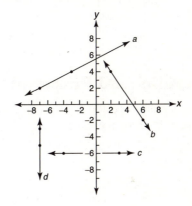

(a) Line a passes through the points $(-7, 2)$ and $(-3, 4)$.

$$m = \frac{4 - 2}{-3 - (-7)}$$

$$m = \frac{2}{4}$$

$$m = \frac{1}{2}$$

(b) Line b passes through the points $(2, 4)$ and $(6, -2)$.

$$m = \frac{-2 - 4}{6 - 2}$$

$$m = \frac{-6}{4}$$

$$m = -\frac{3}{2}$$

(c) Line *c* is parallel to the *x*-axis. Therefore, *m* = 0.

(d) Line *d* is parallel to the *y*-axis. Therefore, line *d* has an undefined slope.

Example 9: A line passes through (−5, 8) with a slope of 2/3. If another point on this line has coordinates (*x*, 12), find *x*.

$$m = \frac{y_2 - y_1}{x_2 - x_1}$$

$$\frac{2}{3} = \frac{12 - 8}{x - (-5)}$$

$$\frac{2}{3} = \frac{4}{x + 5}$$

$$2(x + 5) = 4(3) \quad \textit{(Cross-Products Property)}$$

$$2x + 10 = 12$$

$$2x = 2$$

$$x = 1$$

Slopes of Parallel and Perpendicular Lines

If lines are parallel, they slant in exactly the same direction. If they are non-vertical, their steepness is exactly the same.

Theorem 103: If two nonvertical lines are parallel, then they have the same slope.

Theorem 104: If two lines have the same slope, then the lines are non-vertical parallel lines.

If two lines are perpendicular and neither one is vertical, then one of the lines has a positive slope, and the other has a negative slope. Also, the absolute values of their slopes are reciprocals.

Theorem 105: If two nonvertical lines are perpendicular, then their slopes are opposite reciprocals of one another, or the product of their slopes is −1.

Theorem 106: If the slopes of two lines are opposite reciprocals of one another, or the product of their slopes is −1, then the lines are nonvertical perpendicular lines.

Horizontal and vertical lines are always perpendicular: therefore, two lines, one of which has a zero slope and the other an undefined slope are perpendicular.

Example 10: If line *l* has slope 3/4, then (a) any line parallel to *l* has slope ___, and (b) any line perpendicular to *l* has slope ___.

(a) 3/4 *(Theorem 103)*

(b) −4/3 *(Theorem 105)*

Example 11: Given points *Q, R, S,* and *T,* tell which sides, if any, of quadrilateral *QRST* in Figure 10-8 are parallel or perpendicular.

$$Q(-1, 0), R(1, 1), S(0, 3), \text{ and } T(-3, 4)$$

Figure 10-8   Determining which sides, if any, of a quadrilateral are parallel or perpendicular.

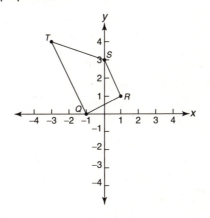

$$m\overline{QR} = \frac{1-0}{1-(-1)} \qquad m\overline{QR} = \frac{1}{2}$$

$$m\overline{RS} = \frac{3-1}{0-1} \qquad m\overline{RS} = -\frac{2}{1}$$

$$m\overline{ST} = \frac{4-3}{-3-0} \qquad m\overline{ST} = -\frac{1}{3}$$

$$m\overline{QT} = \frac{4-0}{-3-(-1)} \qquad m\overline{QT} = -\frac{2}{1}$$

So $\overline{QR} \perp \overline{RS}$ and $\overline{QR} \perp \overline{QT}$ *(Theorem 106)*

and $\overline{RS} \perp \overline{QT}$ *(Theorem 104)*

Equations of Lines

Equations involving one or two variables can be graphed on any *x-y* coordinate plane. In general, the following principles are true:

■ If a point lies on the graph of an equation, then its coordinates make the equation a true statement.

■ If the coordinates of a point make an equation a true statement, then the point lies on the graph of the equation.

A **linear equation** is any equation whose graph is a line. All linear equations can be written in the form $Ax + By = C$ where A, B, and C are real numbers and A and B are not both zero. The following examples are linear equations and their respective A, B, and C values.

$x + y = 0$	$3x - 4y = 9$	$x = -6$	$y = 7$
$A = 1$	$A = 3$	$A = 1$	$A = 0$
$B = 1$	$B = -4$	$B = 0$	$B = 1$
$C = 0$	$C = 9$	$C = -6$	$C = 7$

This form for equations of lines is known as the standard form for the equation of a line.

The **x-intercept** of a graph is the point where the graph intersects the x-axis. It always has a y-coordinate of zero. A horizontal line that is not the x-axis has no x-intercept.

The **y-intercept** of a graph is the point where the graph intersects the y-axis. It always has an x-coordinate of zero. A vertical line that is not the y-axis has no y-intercept.

One way to graph a linear equation is to find solutions by giving a value to one variable and solving the resulting equation for the other variable. A minimum of two points is necessary to graph a linear equation.

Example 12: Draw the graph of $2x + 3y = 12$ by finding the x-intercept and the y-intercept.

The x-intercept has a y-coordinate of zero. Substituting zero for y, the resulting equation is $2x + 3(0) = 12$. Now solving for x,

$$2x = 12$$
$$x = 6$$

The x-intercept is at $(6, 0)$, or the x-intercept value is 6.

The y-intercept has an x-coordinate of zero. Substituting zero for x, the resulting equation is $2(0) + 3y = 12$. Now solving for y,

$$3y = 12$$
$$y = 4$$

The y-intercept is at (0, 4), or the y-intercept value is 4.

The line can now be graphed by graphing these two points and then drawing the line they determine (Figure 10-9).

Figure 10-9 Drawing the graph of a linear equation after finding the x-intercept and the y-intercept.

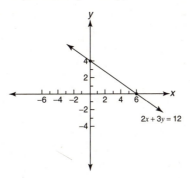

Example 13: Draw the graph of $x = 2$.

$x = 2$ is a vertical line whose x-coordinate is always 2 (Figure 10-10).

Figure 10-10 The graph of a vertical line.

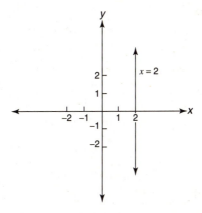

Example 14: Draw the graph of $y = -1$.

$y = -1$ is a horizontal line whose y-coordinate is always -1. See Figure 10-11.

Figure 10-11

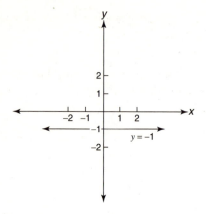

Suppose that A is a particular point called (x_1, y_1) and B is any point called (x, y). Then the slope of the line through A and B is represented by

$$\frac{y - y_1}{x - x_1} = m$$

Applying the *Cross-Products Property*, $y - y_1 = m\,(x - x_1)$. This is the point-slope form of a nonvertical line.

Theorem 107: The point-slope form of a line passing through (x_1, y_1) and having slope m is $y - y_1 = m\,(x - x_1)$.

Example 15: Find the equation of a line containing the points $(-3,4)$ and $(7,2)$ and write the equation in (a) point-slope form and (b) standard form.

(a) For the point-slope form, first find the slope, m.

$$m = \frac{2 - 4}{7 - (-3)}$$

$$m = -\frac{2}{10}$$

$$m = -\frac{1}{5}$$

Now choose either original point—say, $(-3, 4)$.

So, $y - 4 = -1/5(x - -3)$ or $y - 4 = -1/5(x + 3)$

(b) Begin with the point-slope form and clear it of fractions by multiplying both sides by the least common denominator.

$$y - 4 = -1/5(x + 3)$$

Multiply both sides by 5.

$$5(y-4) = 5[-1/5(x+3)]$$
$$5y - 20 = -(x+3)$$
$$5y - 20 = -x - 3$$

Get x and y on one side and the constants on the other side by adding x to both sides and adding 20 to both sides.

$$x + 5y = 17$$

A nonvertical line written in standard form is $Ax + By = C$ with $B \neq 0$. If this equation is solved for y, it becomes

$$By = -Ax + C$$
$$y = -(A/B)x + C/B$$

Let b denote the y-intercept of a line. The point-slope form of the equation of the line passing through $(0, b)$ with slope m is

$$y - b = m(x - 0).$$

Adding b to both sides of the equation yields

$$y = mx + b.$$

This is known as the **slope-intercept form** of the equation of a nonvertical line. Note that, in order to obtain the slope-intercept form, a nonvertical line written in standard form $Ax + By = C$ with $B \neq 0$ can be solved algebraically for y (see Example 16).

Theorem 108: The slope-intercept form of a nonvertical line with slope m and y-intercept value b is $y = mx + b$.

Example 16: Find the slope and y-intercept value of the line with equation $3x - 4y = 20$.

Solve $3x - 4y = 20$ for y.

$$-4y = -3x + 20$$
$$y = (3/4)x - 5$$

Therefore, the slope of the line is 3/4 and the y-intercept value is −5.

Example 17: Line l_1 has equation $2x + 5y = 10$. Line l_2 has equation $4x + 10y = 30$. Line l_3 has equation $15x - 6y = 12$. Which lines, if any, are parallel?

Put each equation into slope-intercept form and determine the slope of each line.

l_1:

$$2x + 5y = 10$$
$$5y = -2x + 10$$
$$y = (-2/5)x + 2$$
$$\text{slope } l_1 = -2/5$$

l_2:

$$4x + 10y = 30$$
$$10y = -4x + 30$$
$$y = (-2/5)x + 3$$
$$\text{slope } l_2 = -2/3$$

l_3:

$$15x - 6y = 12$$
$$-6y = -15x + 12$$
$$y = (5/2)x - 2$$
$$\text{slope } l_3 = 5/2$$

Slope l_1 = slope l_2, therefore; $l_1 \parallel l_2$ by *Theorem 104.*

Because (slope l_1)(slope l_3) = -1 and (slope l_2)(slope l_3) = -1, $l_1 \perp l_3$ and $l_2 \perp l_3$ by *Theorem 106.*

Summary of Coordinate Geometry Formulas

If $A(x_1, y_1)$ and $B(x_2, y_2,)$, then

distance d, from A to $B =$

$$d = \sqrt{(x_2 - x_1)^2 + (y_2 - y_1)^2}$$

midpoint, M, of $\overline{AB} =$

$$M = \left(\frac{x_1 + x_2}{2}, \frac{y_1 + y_2}{2}\right)$$

slope, m, of $\overleftrightarrow{AB} =$

$$m = \frac{y_2 - y_1}{x_2 - x_1}$$

Following is a list of the equations of lines:

Standard form: $Ax + By = C$

A, B, and C are real numbers

A and B are not both zero

Point-slope form: $y - y_1 = m(x - x_1)$

(x_1, y_1) is a point on the line and m is the slope of the line

Slope-intercept form: $y = mx + b$

m is the slope of the line and b is the y-intercept value

Chapter Checkout

Q&A

1. What is the distance from (3,3) to (10,–1)?

2. What is the midpoint of the line segment with endpoints (3,3) and (10, –1)?

3. What is the slope of a line passing through (3,3) and (10, –1)?

4. Find the following forms for the equation of the line passing through (3,3) and (10, –1):

(a) point-slope form
(b) slope-intercept form
(c) standard form

5. (a) Find the slope and the y-intercept of the line with equation $4x + 5y = 6$.

(b) Find an equation of the line passing through (3,3) and perpendicular to the line from (a).

Answers: 1. $\sqrt{65}$ 2. $(\frac{13}{2}, 1)$ 3. $-\frac{4}{7}$ 4. (a) $y - 3 = -\frac{4}{7}(x - 3)$ or $y + 1 = -\frac{4}{7}(x - 10)$ (b) $y = -\frac{4}{7}x + \frac{33}{7}$ (c) $4x + 7y = 33$ 5. (a) $m = -\frac{4}{5}$, $b = \frac{6}{5}$ (b) $y - 3 = \frac{5}{4}(x - 3)$

CQR REVIEW

Use this CQR Review to practice what you've learned in this book and to build your confidence in working with geometry. After you work through the review questions, the critical thinking exercises, and the projects, you're well on your way to achieving your goal of being proficient in the use of geometry.

Chapter 1

1. In the following diagram, determine the degree measure of (a) ∠ *a* and (b) ∠ *b*.

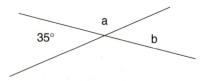

2. If two lines are parallel, then how many points are in their intersection?

3. True or False: Two lines perpendicular to the same line are perpendicular to each other.

4. The undefined terms that were identified as such in *CQR Geometry* are *point*, *line*, and *plane*. What other terms were treated as undefined here, even though they may not have been labeled as such? (Chapter 1)

Chapter 2

5. In the following diagram, two parallel lines are intersected by a transversal. What is the degree measure (a) of ∠ *a*? (b) of ∠ *b*? (c) of ∠ *c*?

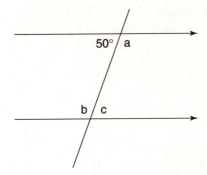

6. In a triangle with two angles of measure 30° and 40°, what must be the measure of the third angle?

7. In the following figure, $m\angle\,a = 45°$ and $m\angle\,d = 145°$. What is the degree measure of $\angle\,f$?

8. Which set of numbers may be the lengths of the sides of a triangle?
(a) {1,3,5} (b) {4,5,6} (c) {4,5,9} (d) {4,5,10}

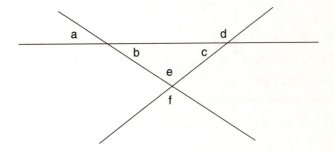

9. In the following diagram, determine the degree measure of $\angle\,a$.

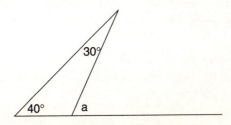

10. Gather together several old rulers and drill holes near the ends of each one. Then bolt three of them together in the shape of a triangle. Test to see that the triangle is rigid in the sense that its shape cannot be distorted without breaking the triangle apart. Now try this with four rulers. Is the quadrilateral you constructed also rigid or can you distort its shape without breaking the figure apart? What about a pentagon? If these figures are not rigid, can you think of a way to make them so? Can you imagine why this kind of question might be important to an engineer? (Chapters 3 and 4)

11. A farmer owns a large piece of property whose boundary is a quadrilateral. He wants to determine the area of this property. He knows that the three sides of a triangle uniquely determine the area of the triangle, and, in fact, he even knows the formula for that area. He believes that this idea generalizes to quadrilaterals, although he does not know a formula for area of quadrilaterals. He has a friend who is a mathematician, so he plans to measure the lengths of the four sides forming the boundaries of his property and take those measurements to his friend. Will these measurements alone be enough information for the mathematician to properly determine the area of the farmer's property? (Hint: Consider a square and a rhombus with edges having the same length.) (Chapters 3, 4, and 5)

Chapter 4

12. True or False: A concave polygon has at least one interior angle with degree measure more than 180°.

13. Draw a quadrilateral that is not convex.

14. What is the degree measure of each interior angle in a regular pentagon?

15. What is the degree measure of an exterior angle of a regular nonagon?

16. Each side of an equilateral triangle has length 6. A second triangle is formed by the segments that join the midpoints of the sides of the equilateral triangle. What is the perimeter of the second triangle?

Chapter 5

17. Compute (a) the circumference and (b) the area of a circle of radius $r = 5$.

18. Determine (a) the perimeter and (b) the area of $\triangle ABC$ in the following diagram.

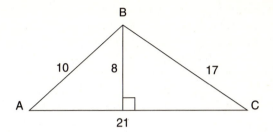

Chapter 6

19. The three angles of a triangle are in the ratio 1 : 2 : 6. What is the measure of the smallest angle?

20. What is the degree measure of the angle formed by the hands of a clock at 4:00?

21. True or False: The area of a square will be doubled if the length of each side is multiplied by $\sqrt{2}$.

Chapter 7

22. If the legs of an isosceles triangle are each 17 inches long and the altitude to the base is 15 inches long, how long is the base of the triangle?

23. In a square, the perimeter is how many times as long as a diagonal?

24. What is the area of an equilateral triangle whose side has length 8?

25. The altitude drawn to the hypotenuse of a right triangle divides the hypotenuse into two segments with lengths 3 and 10. What is the length of the altitude?

26. In the following diagram, *AE* and *BD* are both perpendicular to *EC*. If *BC* = 5 and *BD* = 3, what is the ratio of *AE* to *EC*?

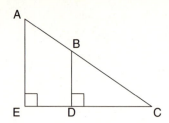

27. True or False: All isosceles right triangles must be similar.

28. Which of the following sets of numbers may be the lengths of the sides of a right triangle? (a) {4, 6, 8} (b) {12, 16, 20} (c) {7, 17, 23} (d) {9, 20, 27}

29. If the hypotenuse of a 30°-60°-90° triangle is 5, what is the length of the side opposite the 60° angle?

30. Jeremy lives on the corner of a rectangular field that measures 120 yards by 160 yards. If he wants to walk to the diagonally opposite corner, he can either go along the boundary of the field or cut across in a straight line. How many yards does he save by taking the direct route?

31. In the following diagram, *ABDE* is a parallelogram and *BCEF* is a square. If *AB* = 10 and *CD* = 6, what is the perimeter of the parallelogram *ABDE*?

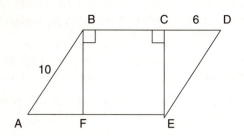

32. In the following rectangle *ACED*, *B* is on $\overline{AC}$ and △*BDE* is equilateral. Find the length of $\overline{AD}$.

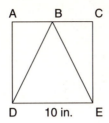

A   B   C

D   10 in.   E

33. The following figure shows a rectangle inscribed in a circle. If the measurements of the rectangle are 8 inches by 14 inches, what is the area of the circle?

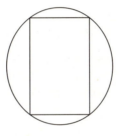

Chapter 8

34. From the same external point, two tangents are drawn to a circle. The tangents intercept arcs whose measures are 300° and 60°. What is the measure of the angle formed by the tangents?

35. The measure of the central angle of a sector is 120°. If the radius of the circle is 6 centimeters, determine the area of the sector.

36. If a central angle of 45° intercepts an arc 5 inches long on a circle, what is the radius of the circle?

37. If an equilateral triangle is inscribed in a circle of radius 1 inch, what will be the length of one of the triangle's sides?

38. What is the measure of $\angle\ a$ in the following figure?

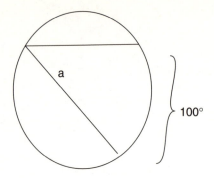

Chapter 9

39. Determine (a) the total surface area and (b) the volume of the rectangular solid in the following figure.

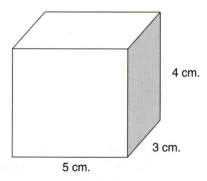

4 cm.

3 cm.

5 cm.

40. Find (a) the lateral area, (b) total area, and (c) volume of a right circular cone with radius 4 inches and height 10 inches.

41. *CQR Geometry* supplies definitions of perpendicular lines and perpendicular planes. No formal definition was given for a line to be perpendicular to a plane, although comfortableness with this notion was assumed in Chapter 9. After looking back at the definition of perpendicular planes, tell what you think would be a reasonable definition for "line l is perpendicular to plane P." (Chapters 1 and 9)

Chapter 10

42. Find (a) the surface area and (b) the volume of a sphere of radius *r* = 7 in.

43. The vertices of the rectangle *ABCD* are *A*(2,0), *B*(5,0), *C*(5,4), and *D*(2,4). How long is the diagonal *AC*?

44. The coordinates of the endpoints of the diameter of a circle are (–1,5) and (3,–1). Specify the coordinates of the center of the circle.

45. True or False: If two points have the same *y*-coordinate, then they lie on the same horizontal line.

46. What is the distance between the points (2,5) and (10,20)?

47. Find the slope-intercept form of the equation of the line passing through (2,5) and (5,9).

48. Go to the physics laboratory at your local high school or college. Ask the instructor to indicate ways that geometry plays a role in various physics experiments.

49. It has been said that the basis of surveying is geometry. Talk to someone who surveys property boundary lines. Ask what parts of geometry are most important in the surveyor's work.

Answers: 1. (a) 145° (b) 35° **2.** 0 **3.** False **4.** Provide your own answer **5.** (a) 130° (b) 130° (c) 50° **6.** 110° **7.** 100° **8.** b **9.** 70° **10.** Provide your own answer **11.** 70° **12.** True

13.

14. 108° **15.** 40° **16.** 9 **17.** (a) 10 π (b) 25 π **18.** (a) 48 (b) 84 **19.** 20° **20.** 120° **21.** True **22.** 16 inches **23.** $2\sqrt{2}$ **24.** $16\sqrt{3}$ **25.** $\sqrt{30}$ **26.** 3 : 4 **27.** True **28.** b **29.** $\frac{5}{2}\sqrt{3}$ **30.** 80 yards **31.** 48 **32.** $5\sqrt{3}$ **33.** 65 π **34.** 120° **35.** 12 π cm^2 **36.** $\frac{20}{\pi}$ inches **37.** $\sqrt{3}$ **38.** 50° **39.** (a) 94 cm^2 (b) 60 cm^3 **40.** (a) $8\sqrt{29}\,\pi$ in^2 (b) $(8\sqrt{29}+16)\,\pi$ in^2 (c) $\frac{160}{3}\,\pi$ in^3 **41.** Provide your own answer **42.** (a) 196 π in^2 (b) $\frac{1372}{3}\,\pi$ in^3 **43.** 5 **44.** (1,2) **45.** True **46.** 17 **47.** y = $\frac{4}{3}$x + $\frac{7}{3}$ **48.** Provide your own answer **49.** Provide your own answer

CQR RESOURCE CENTER

The learning doesn't need to stop here. CQR Resource Center gives you a rundown of the best information in print and online about geometry. You can also find all kinds of pertinent information at www.cliffsnotes.com. Look for all the terrific resources at your favorite bookstore or local library and on the Internet. When you're online, make your first stop www.cliffsnotes.com where you'll find more incredibly useful information about geometry.

Books

This CliffsQuickReview book is one of many great books about geometry. If you want some additional resources, check out these books for more information about geometry and related disciplines.

Geometry Review Guide, by Isidore Dressler, integrates geometry with arithmetic, algebra, and numerical trigonometry while providing more details and many exercises concerning some topics presented in CQR *Geometry.* This book also includes sections on reasoning. AMSCO School Publications, Inc., 1973.

Schaums Outline of Geometry, Third Edition, by Barnett Rich, contains chapters on algebra, proofs of theorems, and the use of the graphing calculator in addition to chapters on the topics presented in CQR *Geometry;* there are many exercises (with answers) throughout the book. McGraw-Hill Companies, Inc., 2000.

Merrill Geometry: Applications and Connections, by Gail F. Burrill et al., is a textbook that presents a few additional topics as well as more details on topics presented in condensed form in CQR *Geometry.* Merrill Publishing Company, Glencoe/McGraw-Hill, 1995.

It's easy to find other mathematics books published by Hungry Minds. You'll find them in your favorite bookstores (on the Internet and at a store near you). We have three Web sites that you can use to read about all the books we publish.

- www.cliffsnotes.com
- www.dummies.com
- www.hungryminds.com

Internet

Check out these Web sites for more information about geometry.

Geometry — `http://library.thinkquest.org/2647/geometry/geometry.htm` provides information on almost everything you'll ever need to know about geometry.

Geometry Online — `http://math.rice.edu/~lanius/Geom/` offers activities for middle and high school geometry.

The University of Texas at Austin — `www.ece.utexas.edu/projects/k12-fall98/14535/Group7/` gives a ton of information about parallel lines and planes. You'll also find various geometry formulas and facts as well as real world examples.

Math for Morons Like Us — `http://library.thinkquest.org/20991/home.html` has been designed to "assist you in your pursuit of increased mathematical understanding, or *whatever sounds good to you.*" The subjects covered range from Pre-Algebra to Calculus. This site will hopefully clarify some of those confusing math concepts. You know, the ones that have been waking you up in the middle of the night for so long! This site has tutorials, sample problems, and quizzes.

Ask a Question — `www.awesomelibrary.org/Office/Main/InvolvingStudents/AskaQuestion.html` offers opportunities to ask experts in all fields questions not just about geometry, although geometry help is one of the features available. You might want to bookmark this Web site for all your homework needs.

Next time you're on the Internet, don't forget to drop by `www.cliffsnotes.com`. We created an online Resource Center that you can use today, tomorrow, and beyond.

Send Us Your Favorite Tips

In your quest for knowledge, have you ever experienced that sublime moment when you figure out a trick that saves time or trouble? If you've discovered a useful tip that helped you understand *CliffsQuickReview Geometry* more effectively and you'd like to share it, the CliffsNotes staff would love to hear from you. Go to our Web site at `www.cliffsnotes.com` and click the Talk to Us button. If we select your tip, we may publish it as part of CliffsNotes Daily, our exciting, free email newsletter. To find out more or to subscribe to a newsletter, go to `www.cliffsnotes.com` on the Web.

Glossary

acute angle—a positive angle whose measure is less than 90°.

acute triangle—a triangle having all acute angles in its interior.

adjacent angles—any two angles that share a common vertex and a common side that separates them.

alternate exterior angles—the nonadjacent angles outside two lines being intersected by a transversal, on opposite sides of the transversal.

alternate interior angles—the nonadjacent angles within two lines being intersected by a transversal, on opposite sides of the transversal.

altitude of a prism—a segment perpendicular to the planes of the bases with an endpoint in each plane.

altitude of a pyramid—the perpendicular line segment from the vertex to the plane of the base.

altitude of a triangle—the perpendicular line segment from a vertex to the side opposite (or its extension).

angle—formed by two rays with the same endpoint (vertex).

angle bisector—a ray that divides an angle into two equal angles.

apothem of a regular polygon—a line segment that goes from the center of the polygon to the midpoint of one of the sides, forming a right angle.

arc of a circle—a connected portion of the circle.

area—a measure of the interior of a planar (flat) figure. It is expressed in square units such as square inches (in^2) or square centimeters (cm^2) or in special units such as acres.

base angles of an isosceles triangle—the two angles other than the vertex angle.

base angles of a trapezoid—a pair of angles sharing the same base.

base of an isosceles triangle—the third side, as distinguished from the other two equal sides.

base of a triangle—any side of the triangle.

bases of a parallelogram—each pair of parallel sides can serve as bases.

bases of a prism—the two congruent polygons lying in parallel planes.

bases of a trapezoid—the pair of parallel sides.

center of a circle (sphere)—the (fixed) interior point that is equidistant from all points on the circle (sphere).

center of a regular polygon—the interior point that is equidistant from each of the vertices.

central angle—an angle formed by any two radii in a circle.

chord—a line segment whose endpoints lie on a circle.

circle—a planar (flat) figure with all of its points equidistant from a fixed point (the center).

circumference—the distance around a circle.

collinear points—points that lie on the same line.

common tangent—a line that is tangent to two circles in the same plane.

complementary angles—two positive angles whose sum is 90°.

concave polygon—a polygon that is not convex. A line segment connecting two vertices may pass outside the figure.

congruent triangles—triangles that have exactly the same size and shape.

consecutive exterior angles—exterior angles on the same side of the transversal.

consecutive interior angles—interior angles on the same side of the transversal.

consecutive sides of a polygon—two sides that have a common endpoint.

convex polygon—a polygon for which any two interior points can be connected by a line segment that stays entirely inside the polygon.

coordinates of a point—the ordered pair of numbers assigned to a point in a plane.

coordinate plane—the x-axis, the y-axis, and all the points in the plane they determine.

corresponding angles—angles that occur in the same relative position in each group of four angles created when a transversal intersects two lines.

corresponding parts of triangles—the parts of two (usually) congruent or similar triangles that are in the same relative positions.

cube—a square right prism whose lateral edges are the same length as a side of the base.

cubic inch—a measure of the interior of a cube whose lateral edge has a length of one inch.

cylinder—a prism-like solid whose bases are circles.

decagon—a ten-sided polygon.

degree—a measure of an angle. It is one three-hundred-and-sixtieth ($\frac{1}{360}$) of a revolution.

degree measure of a major arc—360° minus the degree measure of the minor arc that has the same endpoints as the major arc.

degree measure of a minor arc—the degree measure of the central angle associated with the arc.

degree measure of a semicircle—180°.

diagonal of a polygon—any line segment that joins two nonconsecutive vertices of the polygon.

diameter of a circle—a chord that passes through the center of the circle.

equiangular triangle (polygon)—a triangle (polygon) with all angles equal in measure.

equilateral triangle (polygon)—a triangle (polygon) with all sides equal in measure.

exterior angle of a triangle—the nonstraight angle formed outside the triangle when one of its sides is extended; it is adjacent to an interior angle of the triangle.

exterior angle sum—the sum of the measures of all the exterior angles of a polygon, one angle at each vertex.

external common tangent—a common tangent that does not intersect the segment joining the centers of two circles.

extremes of a proportion—when a proportion is written in the form $a : b = c : d$, a and d are referred to as extremes of the proportion.

geometric mean (or mean proportional)—a positive value that is repeated in either the means or the extremes positions of a proportion.

graph of an ordered pair—the point (in the coordinate plane) associated with an ordered pair of real numbers.

height of a parallelogram (trapezoid)—any perpendicular segment connecting two bases of the parallelogram (trapezoid).

heptagon—a seven-sided polygon (also: septagon).

hexagon—a six-sided polygon.

hypotenuse—the side opposite the right angle in a right triangle.

inscribed angle—an angle formed by two chords with the vertex on the circle.

intercepted arc—the connected portion of a circle that lies in the interior of the intercepting angle, together with the endpoints of the arc.

interior angle sum—the sum of the measures of all the interior angles of a polygon.

internal common tangent—a common tangent that intersects the segment joining the centers of two circles.

intersecting lines—two or more lines that meet in a single point.

isosceles triangle—a triangle in which two sides have equal measures.

isosceles right triangle—a triangle with two equal sides, two equal angles, and a right angle.

isosceles trapezoid—a trapezoid whose two legs are equal.

lateral area of a right prism—the sum of the areas of all the lateral faces.

lateral edges of a prism—the parallel line segments formed by connecting the corresponding vertices of the two base polygons.

lateral edges of a pyramid—the segments where the lateral faces meet.

lateral faces of a prism—the parallelograms formed by the lateral edges.

lateral faces of a pyramid—the triangular sides; they all meet at the vertex.

legs of an isosceles triangle—the two equal sides.

legs of a right triangle—the two sides other than the hypotenuse.

legs of a trapezoid—the nonparallel sides.

line—an undefined term. A line can be visualized as a connected set of infinitely many points extending (without curves) infinitely far in opposite directions.

linear equation—an equation whose graph is a straight line.

line segment—a connected piece of a line with two endpoints.

major arc—an arc that is larger than a semicircle.

means of a proportion—when a proportion is written in the form $a : b = c : d$, b and c are referred to as means of the proportion.

median of a triangle—a line segment drawn from a vertex to the midpoint of the opposite side.

median of a trapezoid—a segment that joins the midpoints of the legs.

midpoint of a line segment—the point on the segment equidistant from the endpoints; the halfway point.

minor arc—an arc that is smaller than a semicircle.

n-gon—a polygon with n sides.

nonagon—a nine-sided polygon.

noncollinear points—points that do not all lie on a single line.

oblique prism—a prism whose lateral edges are not perpendicular to the planes of its bases.

obtuse angle—an angle whose measure is more than 90° but less than 180°.

obtuse triangle—a triangle having an obtuse angle in its interior.

octagon—an eight-sided polygon.

ordered pair—a pair of numbers whose order is important; these are used to locate points in the plane.

origin—in two dimensions, the point $(0,0)$; it is the intersection of the x-axis and the y-axis.

parallel lines—two lines that lie in the same plane and never intersect.

parallel planes—two planes that do not intersect.

parallelogram—any quadrilateral with both pairs of opposite sides parallel.

pentagon—a five-sided polygon.

perimeter—the distance around a figure (polygon).

perpendicular lines—lines that intersect to form right angles.

perpendicular planes—A line l is perpendicular to plane A if l is perpendicular to all of the lines in plane A that intersect l. (Think of a stick standing straight up on a level surface. The stick is perpendicular to all of the lines drawn on the table that pass through the point where the stick is standing.)

A plane B is perpendicular to a plane A if plane B contains a line that is perpendicular to plane A. (Think of a book balanced upright on a level surface.)

plane—an undefined term. A plane can be visualized as a flat surface that extends infinitely far in all directions.

point—an undefined term. A point is represented by a dot.

point of tangency—the point where a tangent line intersects a circle.

point-slope form of the equation of a line—the form $y - y_1 = m(x - x_1)$, where m is the slope of the line and (x_1, y_1) is a specific point on the line.

polygon—a closed planar figure with three or more sides, but not infinitely many.

postulate—a statement that is assumed to be true (without proof).

prism—a type of solid with two congruent polygons lying in parallel planes for bases and with three or more lateral faces.

proportion—an equation stating that two ratios are equal.

pyramid—a solid with one base (a polygon) whose vertices are each joined by line segments to a special point not in the plane of the base; these segments form edges for the lateral faces.

Pythagorean Theorem—in a right triangle, the square of the hypotenuse is equal to the sum of the squares of the other two sides of the triangle.

Pythagorean triple—three positive integers *a*, *b*, and *c* that satisfy the equation

$$a^2 + b^2 = c^2.$$

quadrilateral—a four-sided polygon.

quadrants—the four regions of the coordinate plane separated by the *x*-axis and the *y*-axis.

radius of a circle (sphere)—a line segment with the center of the circle (sphere) and a point on the circle (sphere) as endpoints. (plural: *radii*)

radius of a regular polygon—a segment that goes from the center to any vertex.

ratio of two numbers *a* and *b*—the fraction *a/b*, usually expressed in simplest form. Also denoted a : b.

ray—a connected piece of a line with one endpoint and extending infinitely far in one direction; a "half-line."

rectangle: A quadrilateral in which all the angles are right angles.

regular polygon—a polygon that is both equilateral and equiangular.

regular pyramid—a pyramid whose base is a regular polygon and whose lateral edges are all equal.

rhombus—a quadrilateral with all four sides equal.

right angle—a 90° angle.

right circular cone—similar to a regular pyramid, except that its base is a circle.

right circular cylinder—a cylinder with the property that the segment joining the centers of the circular bases is perpendicular to the planes of the bases.

right prism—a prism whose lateral edges are perpendicular to the planes of its bases.

scalene triangle—a triangle in which all three sides have different lengths.

secant—a line that contains a chord.

sector of a circle—a region bounded by two radii and an arc of the circle intercepted by (an angle formed by) those two radii.

semicircle—an arc whose endpoints are the endpoints of a diameter of the circle.

septagon—a seven-sided polygon (also: heptagon).

sides of an angle—the rays that form the angle.

similar polygons—polygons with the same shape; all their angles are corresponding parts.

slant height of a regular pyramid—the altitude of any of the congruent isosceles triangles that form the lateral faces.

slope-intercept form of the equation of a line—the form $y = mx + b$ where m is the slope of the line and b is the y-intercept.

slope of a line—a measurement of direction and steepness of a nonvertical line. Denoted by m, it is calculated by taking points (x_1, y_1) and (x_2, y_2) on the line and computing $m = \dfrac{y_2 - y_1}{x_2 - x_1}$.

sphere—the set of all points in space that are equidistant from a fixed point (the center).

square—a quadrilateral in which all the angles are right angles and all the sides are equal.

standard form for the equation of a line—the form $Ax + By = C$, where A, B, and C are real numbers and A and B are not both zero.

straight angle—a 180° angle.

supplementary angles—two positive angles whose sum is 180°.

tangent—a line that lies in the same plane as a circle and intersects the circle in exactly one point.

theorem—a statement that can be proved using definitions, postulates, and previously proved theorems.

total area of a right prism—the sum of the lateral area and the areas of the two bases.

transversal—a line that intersects two or more lines in the same plane at different points.

trapezoid—a quadrilateral with only one pair of opposite sides parallel.

triangle—a three-sided figure (polygon).

vertex—(1) the common endpoint for the two rays that determine an angle; (2) an endpoint of a side of a polygon.

vertex angle of an isosceles triangle—the angle formed by the two equal sides.

vertex of a pyramid—the special point, not in the plane of the base, joined to each vertex of the base by a line segment that forms an edge of a lateral face.

vertical angles—any pair of nonadjacent angles formed when two lines intersect.

volume—a measure of the interior of a solid; the number of unit cubes necessary to fill the interior of such a solid.

x-axis—a horizontal line used to help locate points in a plane.

x-coordinate—the first term of an ordered pair; it appears to the left of the comma and indicates movement along the x-axis.

x-intercept—the point where a graph intersects the x-axis.

y-axis—a vertical line used to help locate points in a plane.

y-coordinate—the second term of an ordered pair; it appears to the right of the comma and indicates movement along the y-axis.

y-intercept—the point where a graph intersects the y-axis.

Index

continued

continued